NATURAL DISASTER RESEARCH, PREDICTION AND MITIGATION

EARTHQUAKES

MONITORING TECHNOLOGY, DISASTER MANAGEMENT AND IMPACT ASSESSMENT

NATURAL DISASTER RESEARCH, PREDICTION AND MITIGATION

Additional books in this series can be found on Nova's website under the Series tab.

Additional e-books in this series can be found on Nova's website under the eBooks tab.

NATURAL DISASTER RESEARCH, PREDICTION AND MITIGATION

EARTHQUAKES

MONITORING TECHNOLOGY, DISASTER MANAGEMENT AND IMPACT ASSESSMENT

WAYNE COLEMAN
EDITOR

nova publishers
New York

Library of Congress Cataloging-in-Publication Data

Names: Coleman, Wayne C., editor.
Title: Earthquakes : monitoring technology, disaster management, and impact assessment / Wayne Coleman, editor.
Description: Hauppauge, New York : Nova Science Publishers, Inc., [2016] | Series: Natural disaster research, prediction, and mitigation | Includes index.
Identifiers: LCCN 2016045423 (print) | LCCN 2016047160 (ebook) | ISBN 9781536103427 (hardcover) | ISBN 9781536103564
Subjects: LCSH: Earthquakes. | Earthquake hazard analysis. | Earthquake resistant design.
Classification: LCC QE534.3 .E3745 2016 (print) | LCC QE534.3 (ebook) | DDC 551.22--dc23
LC record available at https://lccn.loc.gov/2016045423

Published by Nova Science Publishers, Inc. † New York

CONTENTS

PREFACE

In recent years, a number of disasters caused by earthquakes have demonstrated the vulnerability of both the developing and the developed world. This book provides new research on earthquakes. Chapter One focuses on the behavior of a simple spring-block model from the viewpoint of nonlinear dynamics and seismology. Chapter Two employs a new technique based on Extreme Learning Machine (ELM) for determination of liquefaction susceptibility of soil based on Standard Penetration Test (SPT) and Cone Penetration Test (CPT) from the Chi-Chi earthquake. Chapter Three presents a review of earthquake phenomenon in Nigeria, the occurrence and available data. Chapter Four describes the behavior and failure mechanisms of unreinforced masonry buildings and different rehabilitation and strengthening techniques. Chapter Five presents the results related to the preferred sources of information related to Risk, Hazard and Natural disaster. Chapter Six presents results regarding urban and semi-rural schoolchildren's seismic risk perception, awareness and preparedness of a highly seismic region, in the state of Oaxaca, Mexico. Chapter Seven reviews the anomalous decrease in groundwater radon before the Taiwan large earthquakes.

Chapter 1 – Present chapter is focused on the behavior of a simple spring-block model from the viewpoint of nonlinear dynamics and seismology. In particular, authors briefly describe the previously applied methods of nonlinear dynamics for the analysis of the motion of several different models: mono-block model, two-block model and models with an array of blocks with global coupling and varying neighboring interactions, by assuming the rate dependent and rate-and state-dependent friction law. Dynamics of such models is examined under the impact of frictional memory effect, delayed and varying interaction among the blocks, transient and continuous perturbations and random seismic noise. Results of the previous studies show that irregular motion of the examined models, as an analog of the seismogenic fault motion, could be identified as an example of deterministic chaos, or it could occur due to effect of the introduced seismic noise or the existence of global attractor. In such regime, it is confirmed that recorded displacements of spring-block models obey Gutenberg-Richter power law distribution. In case when only the first Hopf bifurcation is detected, proposed models could be used as reliable characteristic earthquake models. Phase response curves of fault dynamics indicated the following: (1) for a mono-block fault model, single pulse perturbation delays the oscillation cycle phase, while two successive pulses induce two different mechanisms of phase advancement; (2) for a homogeneous two-block fault model, when each block is affected by a single pulse perturbation, both blocks exhibit the advance of oscillation phase, (3) for a heterogeneous two-block fault model, a single pulse perturbation has a destabilizing effect of perturbation via interaction between the blocks is non-negligible and its influence of non-perturbed block is destabilizing. Results of previous studies presented in this chapter confirm the reliability of the use of simple spring-block models as phenomenological models of seismogenic fault motion.

Chapter 2 – Seismic liquefaction is a major concern for any earthquake prone area. This article employs a new technique based on Extreme Learning Machine (ELM) for determination of liquefaction susceptibility of soil based on Standard Penetration Test (SPT) and Cone Penetration Test (CPT) from the Chi-Chi earthquake. ELM is the modified version of Single-hidden Layer Feed forward Neural Network. Different input combinations have been tried to get best performance. A comparative study has been carried out between the SPT and CPT based models. CPT based models give better performance than the SPT based model. The developed ELM model shows that Cone Resistance (q_c) and Peck Ground Acceleration (PGA) are sufficient parameters for determination of liquefaction susceptibility of soil.

Chapter 3 – This chapter presents a review of earthquake phenomenon in Nigeria, the occurrence and available data. The review also includes recommendation to government agencies and designers of Engineering Structures on the need to accommodate mitigating efforts to avoid possible future disasters occasioned by earthquake occurrence. Nigeria may not be presently in high impact zone of earthquake phenomenon, but what the future portends cannot be easily predicted. Hence there is necessity for long term planning to avoid being caught unaware of the possible effect of sudden high impact earthquake in its shores. Thus, the need for the application of cutting edge Information Technology and adoption of mitigation techniques in the reduction of earthquake effects cannot be overemphasized. Environmental impact assessment prior to construction should indicate the seismic performance of concrete frames, perhaps under simulated earthquake excitations. Also significant is the effect of rapid urbanization and consequent construction of heavy engineering structures that require deep foundation in coastal cities of Nigeria. The effect of such construction activities can trigger earth movement in the area. Thus, the destructive impacts of earthquake, both on lives and properties could be ameliorated by taking mitigation measures of risks connected to earthquake. Mainly, through adherence to effective town planning and land use plans, enforcement of standard building codes and good construction practices. Moreover, adequate early warning systems, awareness creation and community preparedness would serve as protective mechanism. Finally, hazard and risk mapping and analyses should be carried out on a national and regional basis, so as to enhance emergency preparedness, rescue operation planning and resources allocation.

Chapter 4 – The most common construction techniques for heritage buildings were mainly based on basic geometric rules about stability and earthquake damage observations. This situation led to the construction of massive structures with structural elements of great thickness, but still with poor seismic performance because damage was continuously presented. These restrictions have made historical buildings extremely vulnerable against earthquakes, presenting partial and total collapses. Thus, becomes evident the need of suitable technical solutions to protect them. This chapter describes the behavior and failure mechanisms of unreinforced masonry buildings and different rehabilitation and strengthening techniques. Both techniques are aimed at keeping good deformation compatibility between materials without affecting the architectonic value of the monument. Advantages and disadvantages of different rehabilitation and strengthening proposals are discussed from a sensitized structural engineering perspective taking into account that this type of buildings are very different from modern structures. These techniques aim at protecting cultural heritage against earthquakes by inducing a modified failure mechanisms that involves more energy dissipation than the original structural system.

Chapter 5 – The chapter presents the results related to the preferred sources of information related to Risk, Hazard and Natural disaster. Also, some results of seismic risk perception and the preferred sources of education on seismic risk are presented. The sample size of the cross-sectional study was N = 1489 that included six Preparatory school located in Mexico City. The main findings of the study have been the following: a) overall, the results show that the students are well aware of the words 'Hazard' (99.6%, Mean = 2.00, SD = 0.063), 'Risk' (99.5%, Mean = 2.00, SD = 0.069), and 'Natural disaster' (99.2%, Mean = 2.00, SD = 0.118); b) the main source of information associated with the term 'Risk' (Mean = 3.27, SD = 0.891) was the "School." The term 'Natural disaster', on the other hand, has been found to be "TV" (Mean = 1.87, SD = 0.334); c) regarding the participants' knowledge on the three seismic zones of the Capital City, the overwhelming majority of them did not know this fact (89.1%, Mean = 1.10, SD = 0.294); d) students perceived that an earthquake could occur "At any time" (86.9%, Mean = 4.72, SD = 0.816); e) students prefer to learn more on seismic risk by the following sources and in this particular order: "Civil protection" (44%), "School" (22%), and "Internet" (20%); f) In general, the participants of the study showed a lack of knowledge on seismic risk (see c) above), even though, they are well aware that an earthquake could happen at any time (see d) above). It is hoped that the preliminary results presented here may serve as the basis for a more in-depth analysis, for example, of mass media, social networks related to earthquakes, in the context of the Mexican culture. The paper gives an account of an ongoing research project.

Chapter 6 – The chapter presents results regarding urban and semi-rural schoolchildren's seismic risk perception, awareness and preparedness of a highly seismic region, in the state of Oaxaca, Mexico. The sample of the study was N = 819 (Urban sample size = 522; Semi-rural sample size = 297). The main findings of the cross-sectional study have been the following: a) overall, urban schoolchildren demonstrated a generalized awareness of the seismic area where they live; however, this was not the case for semi-rural schoolchildren; educational campaigns may be needed to further enhance seismic awareness of both urban and semi-rural areas of the region; b) the two hazards associated with earthquakes and crime have been perceived to represent the two major threats to them at school and at home, and for both urban and semi-rural schoolchildren; c) schoolchildren emotions are not consistent when asked whether they would be able to control themselves during an earthquake; d) seismic risk communication takes place at school and at home; this may be regarded as encouraging in order to further educate children on seismic risk; however, more is needed to better educate them on important aspects regarding what to do before, during and after an earthquake and this may well prove to be an effective way to share their knowledge with their parents (most of them are not well educated) and contribute to built a resilient community to earthquakes; e) the preliminary findings of the study are consistent with has been found in similar studies; i.e., "children were able to identify higher frequency events at a higher frequency and lower frequency events at a relatively lower frequency."

Chapter 7 – In southwestern Taiwan, radon anomalous declines prior to the 2010 Mw 6.3 Jiasian and 2016 Mw 6.4 Meinong earthquakes were recorded at the Paihe spring. The above-mentioned earthquakes are the two strongest since 1964 that have occurred near the area southeast of Chiayi. The Paihe spring (P1) is located 46 km and 45 km, respectively, from the epicenters of the 2010 Mw 6.3 Jiasian and 2016 Mw 6.4 Meinong earthquakes. Specifically, the concentration of groundwater radon decreased from background levels of 144 ± 7 and 137 ± 8 pCi/L to minima of 104 ± 8 and 97 ± 9 pCi/L prior to the 2010 Mw 6.3 Jiasian and 2016 Mw 6.4 Meinong earthquakes, respectively. The Paihe spring is from an unconfined limestone aquifer surrounded by ductile shale and sandy shale. The v-shaped pattern of radon behavior in the Paihe spring is quite similar to those radon decreases that were measured in the confined andesite aquifer at Antung in eastern Taiwan.

Recurrent anomalous declines in the concentration of groundwater radon were observed at well D1 in the Antung hot spring in eastern Taiwan prior to the five major earthquakes –2003 Mw 6.8 Chengkung, 2006 Mw 6.1 and Mw 5.9 Taitung, 2008 Mw 5.4 Antung, and 2011 Mw 5.0 Chimei. Well D1 is located 24 km, 52 km, 47 km, 13 km, and 32 km, respectively, from the epicenters of the 2003 Mw 6.8, 2006 Mw 6.1 and Mw 5.9, 2008 Mw 5.4, and 2011 Mw 5.0 earthquakes. The 2003 Mw 6.8 Chengkung was the strongest earthquake near the Chengkung area in eastern Taiwan since 1951. Specifically, radon decreased from a background level of 787 ± 42 pCi/L to a minimum of 326 ± 9 pCi/L prior to the 2003 Mw 6.8 Chengkung earthquake. In a brittle aquifer under un-drained conditions (Antung and Paihe), the dilation of brittle rock mass occurred at a rate faster than the recharge of pore water and gas saturation developed in newly created cracks preceding the earthquake. Radon partitioning into the gas phase may explain the anomalous decrease of groundwater radon prior to the 2003 Chengkung earthquake.

Both observations at Paihe and Antung suggest that in-situ radon volatilization offers an attractive mechanism for a premonitory decrease in groundwater radon. "A low-porosity brittle aquifer with un-drained conditions" is a suitable geological site to consistently detect precursory declines in groundwater radon prior to local large earthquakes. A long-term monitoring of precursory declines in groundwater radon at a suitable geological site can be a useful means for forecasting local disastrous earthquakes. At a suitable geological site, it is of a great practical value to develop the correlations among the scale of radon decline, precursory time, and earthquake magnitude for a given seismogenic fault.

In: Earthquakes
Editor: Wayne Coleman

ISBN: 978-1-53610-342-7

Chapter 1

NONLINEAR DYNAMICS BEHIND THE SEISMOGENIC FAULT MOTION: A REVIEW ON DYNAMICS OF SINGLE-ARRAY SPRING-BLOCK MODELS

Srđan D. Kostić*[1,*], *PhD*, *Nebojša Vasović*[2,†], *PhD*, *Kristina Todorović*[3,‡], *PhD*, *and Igor Franović*[4,§], *PhD

[1]Institute for Development of Water Resources "Jaroslav Černi," Belgrade, Serbia
[2]University of Belgrade Faculty of Mining and Geology, Belgrade, Serbia
[3]University of Belgrade Faculty of Pharmacy, Belgrade, Serbia
[4]University of Belgrade Institute of Physics, Belgrade, Serbia

ABSTRACT

Present chapter is focused on the behavior of a simple spring-block model from the viewpoint of nonlinear dynamics and seismology. In particular, authors briefly describe the previously applied methods of nonlinear dynamics for the analysis of the motion of several different models: mono-block model, two-block model and models with an array

* srdjan.kostic@jcerni.co.rs.
† nebojsa.vasovic@rgf.bg.ac.rs.
‡ kisi@pharmacy.bg.ac.rs.
§ franovic@ipb.ac.rs.

of blocks with global coupling and varying neighboring interactions, by assuming the rate dependent and rate-and state-dependent friction law. Dynamics of such models is examined under the impact of frictional memory effect, delayed and varying interaction among the blocks, transient and continuous perturbations and random seismic noise. Results of the previous studies show that irregular motion of the examined models, as an analog of the seismogenic fault motion, could be identified as an example of deterministic chaos, or it could occur due to effect of the introduced seismic noise or the existence of global attractor. In such regime, it is confirmed that recorded displacements of spring-block models obey Gutenberg-Richter power law distribution. In case when only the first Hopf bifurcation is detected, proposed models could be used as reliable characteristic earthquake models. Phase response curves of fault dynamics indicated the following: (1) for a mono-block fault model, single pulse perturbation delays the oscillation cycle phase, while two successive pulses induce two different mechanisms of phase advancement; (2) for a homogeneous two-block fault model, when each block is affected by a single pulse perturbation, both blocks exhibit the advance of oscillation phase, (3) for a heterogeneous two-block fault model, a single pulse perturbation has a destabilizing effect of perturbation via interaction between the blocks is non-negligible and its influence of non-perturbed block is destabilizing. Results of previous studies presented in this chapter confirm the reliability of the use of simple spring-block models as phenomenological models of seismogenic fault motion.

Keywords: spring-block models, bifurcation analysis, random noise, mean-field approximation, deterministic chaos, power law, time delay, transient perturbation, phase response curve

1. INTRODUCTION

In present chapter, nonlinear dynamics in the background of the seismogenic fault motion is explained through the analysis of motion of spring-block models in the following form:

- mono-block model;
- two-block model;
- one-dimensional array of blocks,

with the assumed rate-dependent and rate-and state-dependent friction law. Since the analyzed models are one-dimensional, present analysis is restricted only to the temporal irregularities. Dynamics of these models is analyzed under the influence of:

- friction memory effect (time delay in friction term);
- delayed interaction among the blocks;
- transient and continuous perturbations;
- random seismic noise;
- varying neighboring interactions.

In the first part of the chapter, dynamics of the mono-block model with the assumed Dieterich-Ruina rate-and state-dependent friction is examined under the impact of friction memory effect, continuous and transient perturbations. Friction memory effect is modeled by assuming by that friction is a function of velocity with a time delay, which was previously shown to be in the range 3-9 ms in laboratory conditions [1], depending on the viscosity and contact load. Standard local bifurcation analysis, demonstrates that the mono-block model with the introduced time delay in the friction term exhibits Ruelle-Takens-Newhouse route to chaos. Results of the performed analysis reveal that small perturbations of the control parameters lead to deterministic chaos, which could be treated as the onset of instability. On the other hand, if regular periodic change of the control parameters is assumed, revealing the effect of the continuous external signal, such as the work of the mining machines, or the traffic impact, results of the performed analysis indicate that deterministic chaos may emerge when some of the parameters are assumed to undergo small oscillations near the equilibrium values, but admitting the periodic motion. In particular, the Ruelle-Takens-Newhouse route to chaos is observed for the cases of limit amplitude perturbations. On the other side, when the angular frequency is assumed constant for the value near the periodic motion of the block in an unperturbed case, variation of oscillation amplitudes probably gives rise to global bifurcations, with immediate occurrence of chaotic behavior. When two perturbed parameters are assumed, further analysis shows that a transition from periodic motion to chaotic behavior occurs for oscillation frequencies of periodic motion in an unperturbed case. Regarding the influence of transient perturbations, the effect of transient external force induced by a passing seismic wave on fault motion is introduced as a change of block acceleration in the form of a sine wave scaled by an exponential pulse. Results obtained indicate the transition to quasiperiodic - or chaos-like motion, which could be attributed to the precursory creep regime and seismic motion, respectively, when fault motion is exposed to external force in the inter-seismic creep regime.

One should note that the introduction of time delay in order to mimic the memory effect, as well as the effect of noise on dynamics of natural systems, although new for theoretical computational geophysics, has already been successfully applied in the area of computational neuroscience [2-14]. Indeed, a very active relationship has been develop between the theory of nonlinear dynamics and neuroscience in the last 20 years, which resemble of positive recurrent feedback. On one hand, application of methods of nonlinear dynamics has substantially contributed to the physical modelling of neureal systems. On the other hand, features detected in neural systems has motivated the development of new concepts in theory of nonlinear dynamics, including excitability, different forms of synchronization and chaotic itinerancy.

In the second part of the chapter, an array of blocks is examined under the impact of the seismic noise, varying neighboring interactions and delayed interactions among the blocks. In particular, an array of 100 blocks is examined with two different types of interaction: (1) all-to-all interaction, (2) varying number of block coupling with 2K neighboring units ($1 \leq 2K \leq N$, $N = 100$). In case of global coupling, results obtained indicate there are local bifurcations from equilibrium state to periodic oscillations, with an occurrence of irregular aperiodic behavior when initial conditions are set away from the equilibrium point. Such a behavior indicates a possible existence of a bi-stable dynamical regime, due to effect of the introduced seismic noise or the existence of global attractor. In this bi-stable regime, distribution of event magnitudes follows Gutenberg-Richter power law with satisfying statistical accuracy, including the b-value within the real observed range. In case when each block is coupled to a varying number of 2K neighboring units, results of the performed analysis indicate the occurrence of direct supercritical Hopf bifurcation, implying a transition from equilibrium state to periodic oscillations under the variation of coupling delay. For loosely coupled blocks and low values of friction, observed system does not exhibit any bifurcation, regardless of the assumed noise amplitude in the expected range of values. Distribution of event magnitudes, as a sum of squared displacements, is found to correspond well to periodic (characteristic) earthquake model.

In the third part of the chapter, the qualitative analysis is focused on determining the first- and second-order phase response curves for models of a simple monoblock fault and paradigmatic two-block complex faults, which display relaxation oscillations in the stick-slip regime [15-18]. Given their strong nonlinear character and the stiffness property, the considered models of earthquake faults are expected to display a number of intricate behaviors, such that their response to perturbation may qualitatively depend on different system parameters, as well as the fault complexity. The main corpus of issues we address includes (i) the sensitivity of a simple monoblock fault to external perturbation, (ii) the influence of system parameters on the profile of phase response curves, (iii) the effect of two consecutive pulse perturbations, as well as (iv) the responses of complex faults, either homogeneous or heterogeneous, to external perturbation. Apart from considering the first-order phase response curves, our interest also lies with the second-order phase response curves because their nontrivial behavior may indicate a potential long-term effect of external perturbation on the duration of an earthquake cycle.

It should be emphasized that results of previous studies which are briefly reviewed in this chapter primarily refer to the description of dynamics of strike-slip fault motion, since the effect of gravity is not taken into account. Nevertheless, the presented methods and approach could be also applied for the analysis of induced seismicity, primarily under the effect of deep underground excavation. In this case, influence of continuous and transient signals on fault motion could be ascribed to different triggering factors, including the effect of blasting and/or drilling or the impact of the vibrations induced by the work of the mining machines.

Present chapter is composed of three different sections. Authors provide brief review of the results of previous studies in Section 2. Dynamics of a mono-block model is given in Section 3, including the description of original model and its derivatives. The main focus of this section will be on frictional memory effect and on the impact of continuous and transient perturbations [19-20]. Dynamics of a chain of 100 blocks is analyzed in Section 4, where the main focus is on the effect of seismic noise, varying neighbor interactions and delayed transition of motion among the blocks [21-22]. Section 5 is devoted to analysis of the external perturbation on earthquake fault dynamics using phase response curves. In Subsection 5.1 we consider the case of a monoblock fault, first analyzing the impact of a single and two successive pulse perturbations, and then demonstrating how the profile of phase response curves depends on the fault parameters. Subsection 5.2 concerns the behavior of the homogeneous and the heterogeneous two-block complex faults. For the homogeneous fault, it is determined how the system responds if each of the blocks is perturbed, but the perturbations arrive with a certain phase lag. For the heterogeneous fault, it is examined how the system response changes depending on whether the block with the shorter or the longer oscillation period is perturbed.

2. Background – Literature Review

It is commonly considered that earthquakes occur as a result of the sudden release of energy within some confined region of undergoing deformation. At the contact of two tectonic plates (interplate earthquakes), or two different lithological units (intraplate earthquakes), energy is stored in the rocks in a form of elastic strain. Faulting occurs when the accumulated strain exceeds the strength of the rock, after which the opposite sides of the fault rebound to equilibrium state (Figure 1). During this process, a part of the stored energy is usually released in a form of heat, other part is used for rock deformation, while the rest of energy propagates as seismic waves. This explanation of earthquake nucleation is known as the elastic rebound theory, formulated by Reid [23] as a result of his studies of the geodetic measurements along the San Andreas fault during the San Francisco earthquake in 1906.

However, even though Reid's hypothesis is quite generally accepted, the main problem arises when the earthquakes occur along the existing faults (reactivated faults). Recent evidence indicates the existence of an aseismic creep along the fault, i.e., the motion along the fault exists without the occurrence of an earthquake. This kind of motion is not predicted by Reid's rebound theory, so there has to be another mechanism that may account for such a motion. One plausible explanation is that any variation of frictional resistance during sliding could induce a dynamic instability, which often occurs repetitively – the instability is followed by a period of no motion during which the stress is recharged, followed by another instability. This common frictional behavior is called the regular stick slip [24].

A stick-slip with friction at the contact of blocks of rocks is considered by many researchers as one of the most simplified and probable analogs of the earthquake source. It is well-known that the major portion of earthquakes occurs in accordance with the scenario corresponding to the model of unstable stick-slip [25].

The suggestion of regular stick-slip motion as a possible mechanism of motion along the fault originally came from Brace and Byerlee [26], who studied the behavior of the granite samples in triaxial shear apparatus. The primary significance of the Brace and Byerlee paper is in the emphasis on stability and the effect of friction, not strength, in explaining the mechanism of earthquakes.

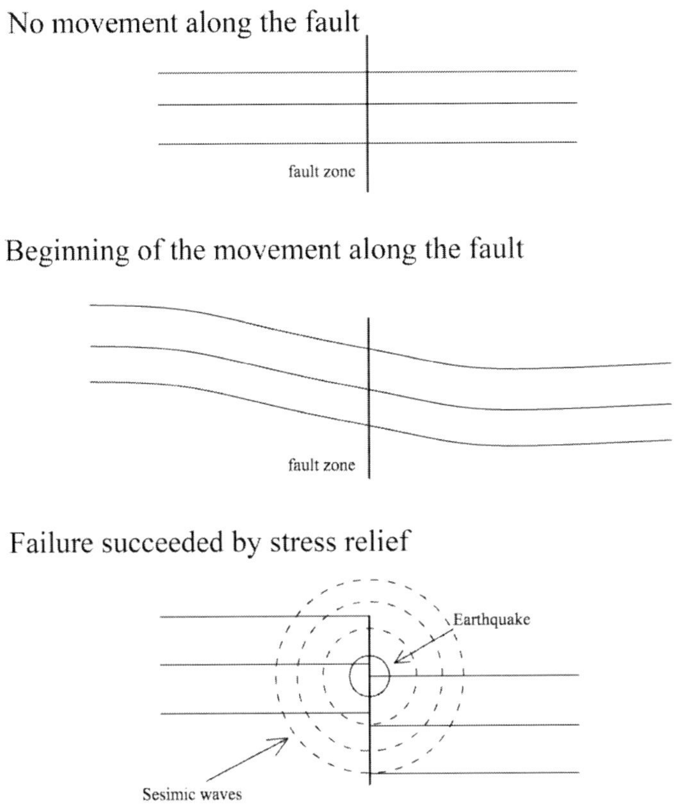

Figure 1. Earthquake nucleation according to Reid's elastic rebound theory.

On the basis of the suggestion by Brace and Byerlee [26], Burridge and Knopoff [27] set an array of blocks interconnected with elastic springs, which are driven along the rough surface by the driving plate. Connection of blocks with the driving plate is also provided by the series of elastic springs. Thereby, they assumed that the main nonlinearity in the analyzed system comes from the specific friction law between the moving block and the rough surface of the lower plate. In this paper, Burridge and Knopoff assumed velocity-dependent friction law of the following form:

$$F = -E\dot{x} + F(\dot{x}) \tag{1}$$

where the first term represents the force due to radiation and the second term combines the viscous and friction effects.

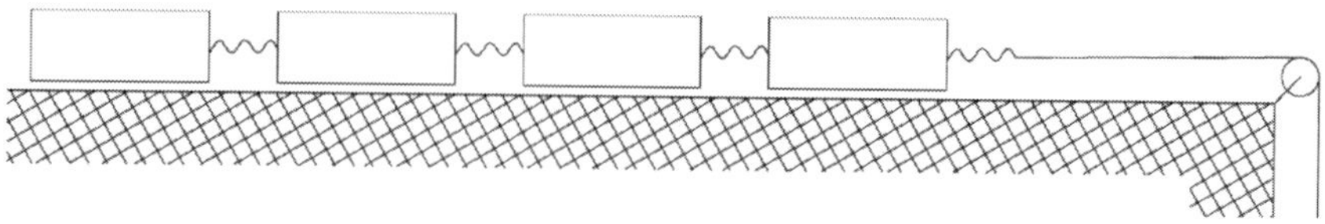

Figure 2. Sketch of the original Burridge-Knopoff model.

Following this suggestion of Brace and Byerlee [26] that stick-slip occurring in laboratory sliding experiments may be analogous to the mechanism of crustal earthquakes, Burridge and Knopoff [27] simulated this stick-slip motion in laboratory conditions, using an array of several blocks. Eight masses and eight springs were placed upon a long, horizontal, plane surface (Figure 2). Approximately uniform friction existed everywhere between the masses and the surface.

Burridge and Knopoff [27] measured the potential energy in the system released during any shock, by assuming Hookean elasticity for all the springs.

The results for the case when all springs are equal show that the parameter values of Gutenberg-Richter law are a = 2.76, b = 0.42, with good approximation (R^2 = 0.996), which are the usual values of parameters for the most observed earthquakes.

In order to compare these values with the real observed data, Kostić et al. [28] conducted a statistical analysis of 483 earthquakes in Serbia with the magnitude range 1.8-5.4, for the period 10/11/2004 - 03/10/2011, according to the data of European-Mediterranean Seismological Centre [28]. The results obtained indicate that the earthquake distribution for the territory of Serbia also follows Gutenberg-Richter law (R^2 = 0,86) with parameters a = 2,64, and b = 0,46 (Figure 3).

In the same paper, Burridge and Knopoff [27] observed the occurrence of aftershocks, relating it to the widely know Omori-Utsu law [29-30]. For this purpose, they observed an array of ten blocks, interconnected by harmonic springs, and also attached through springs to a driving upper plate, which cause the whole system to move in a stick-slip fashion along the rough surface of the lower plate. The histogram for the sequence of 69 aftershocks is shown in Figure 4. Analysis performed by Kostić et al. [28] indicated that the values of parameters of Omori-Utsu law are c = 0.068, and d = 0.012, with the correlation coefficient R^2 = 0,916.

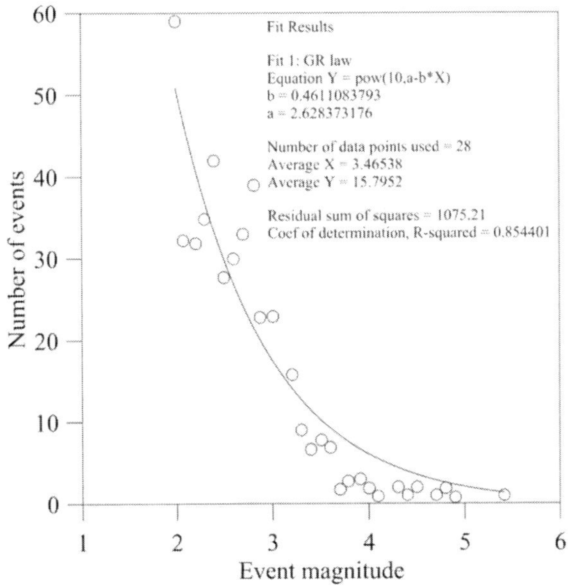

Figure 3. Gutenberg-Richter law for earthquakes in Serbia recorded in the period 2004-2011.

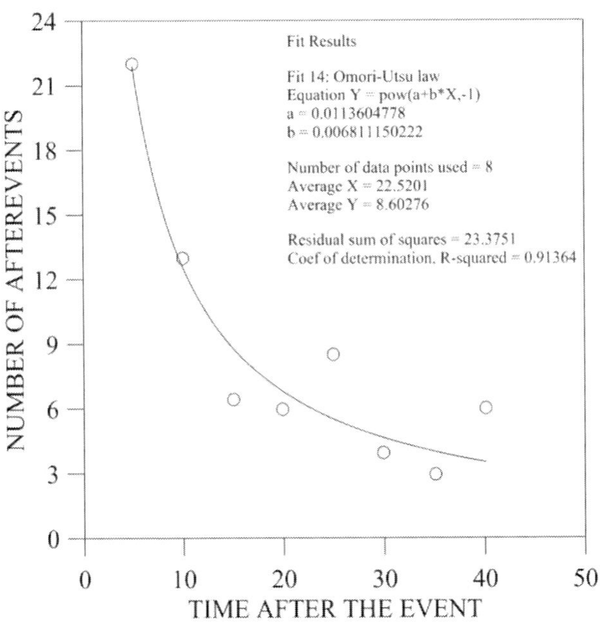

Figure 4. Histogram for number of aftershocks in a time interval of $5x10^5$ units.

Kostić et al. [28] also conducted a statistical analysis of aftershock sequence from 03/11 to 25/11/2011 was done, for the shock following the Kraljevo M5.4 earthquake recorded on November 3rd in 2010 in Serbia. As it can be seen in Figure 5, aftershock sequence of 91 events obey Omori-Utsu law, with the following parameters:

$$N = 1 / (-0{,}06 + 0{,}015t) \qquad (2)$$

with the correlation coefficient $R^2 = 0{,}98$.

The main contribution of the work of Burridge and Knopoff is two-fold:

- they proposed an experimental setup that mimics the stick-slip behavior, and that phenomenologically resembles the motion along the seismogenic fault;
- they showed that the distribution of bursts of motion follows the two significant macroseismological laws: Gutenberg-Richter and Omori-Utsu law.

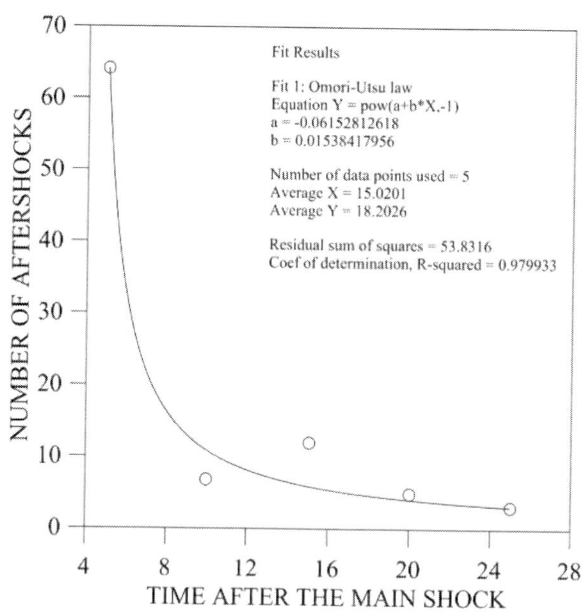

Figure 5. Omori law for aftershock sequence of 2010 Kraljevo M5.8 earthquake in Serbia.

Since the work of Burridge and Knopoff [27], spring-block models have been widely recognized as convenient phenomenological models of fault motion. In particular, a setup consisting of interconnected blocks moving along a rough surface was shown to exhibit stick-slip dynamics, which resembles to real observed movement along the seismogenic faults. Such models have been widely used to mimic the fault motion, in order to capture the main dynamical mechanism behind the fault dynamics. For a very nice review on the existing spring-block model and their main features from the viewpoint of statistical physics, one should refer to the work of Kawamura et al. [31].

After the work of Burridge and Knopoff [27], the subsequent significant study on this topic was performed by Carlson and Langer [32], who analyzed dynamics of a Burridge-Knopoff spring-block model with up to 400 blocks in a row, and by assuming the velocity-dependent friction law:

$$F\left(\dot{x}\right) = F_0 \phi\left(\frac{\dot{X}}{v_1}\right) \tag{3}$$

where φ vanishes at large values of its argument and is normalized so that φ(0) = -φ'(0) = 1, and v_l is some speed that characterizes the velocity dependence of F. Results of their analysis indicate occurrence of three distinct events: microscopic events (including only one or a few blocks), localized events (when triggering pulse excites only the blocks within the starting slipping zone) and delocalized events (when triggering pulse excites the blocks outside the starting slipping zone). With the transition from microscopic to delocalized events, the average velocity increases (approaching unity for delocalized events), and also the event irregularities are being amplified. From the seismological viewpoint, Carlson and Langer provided the relationship between the number of displaced blocks and the frequency of events, as an analogous version of the Gutenberg-Richter power law. In particular, they provided several characteristic magnitude vs. frequency diagrams for different values of control parameters and observed the following. Large number of microscopic events falls at the small end of the magnitude distribution, while localized events, as moderately large events, appear at the center of these diagrams. Distribution of these localized events corresponds well with the Gutenberg-Richter law. In fact, for the value of critical parameter $\alpha = 2.5$, parameter $b = 1$, which is in agreement with the Gutenberg-Richter law. At the large end of the magnitude distribution, delocalized great events appear, which are too frequent to be consistent with the Gutenberg-Richter power law. Results of their research also indicated that dimensionless parameter α (as a rate at which disturbances are amplified as a result of velocity-weakening friction) and fault length have a significant impact on the role of delocalized great events. In particular, great events have a predominant influence in moving the system forward on average at the loading speed and are more nearly periodic on shorter faults and for larger values of α.

Another important paper dealing with the spring-block models from the viewpoint of nonlinear dynamics comes from Vieira [33]. In her paper, Vieira analyzed two-block and three-block model, by assuming also the velocity depending friction law, without the possibility of the backward motion. Results of her research indicated that for a certain parameter values, two-block model exhibits periodic, quasiperiodic and chaotic behavior, with the route to chaos from period one to period two and then directly into chaos. Also, Vieira showed the chaotic synchronization could occur in a three-block model, which could have a significant impact on earthquake prediction, since synchronization decreases the dimensionality of the attractor.

Among the recent research, one should single out the work by Erickson et al. [34-35], who studied the dynamics of a mono-block and a three-block model [34] and model with an array of blocks [35] by assuming the Dieterich-Ruina rate-and state-dependent friction law [36, 37]:

$$\mu(V,\theta) = \mu_0 + A\ln\left(\frac{V}{V_0} + B\ln\left(\frac{V_0\theta}{D_c}\right)\right) \tag{4}$$

where μ_0 is a friction constant for steady-state slip at velocity V_0, V is the velocity of the block, D_c (L) is the critical length, that one block needs to move in order to replace the asperity contact, θ is a state variable, and A and B are empirical constants. State variable θ secures the fading memory of previous states [38]. It represents the delayed reaction of the friction to instantaneous change of the velocity. Material constants A and B strongly depends on the temperature and stress state, and, consequently, they change their values with the depth of the Earth's crust. In particular, parameter (A-B) actually denotes the conditions for the earthquake to occur along the crustal faults. Depending on the sign of this parameter, conditions from the ground surface up to 50km depth change from stable through conditionally stable to unstable, where unstable conditions correspond to the possibility of earthquake occurrence. Indeed, the majority of the recorded earthquake foci is located within the unstable zone, where (A-B) parameter is with the negative sign [39]. When parameter (A-B) is with the negative sign, then one talks about the velocity-weakening law, while positive (A-B) indicates velocity-strengthening law. In order for instability to occur, observed model should exhibit velocity-weakening law. It is interesting to notice that similar law is also observed for the "weak" rock masses (clays), which further implies that such law could be also significant for the landslide activation and its dynamics.

In order to model the details of friction arising from perturbations in average contact lifetime (state) or slip velocity, equation (4) must be coupled with a description of state evolution:

$$\dot{\theta} = 1 - \left(\frac{V\theta}{D_c} \right) \tag{5}$$

Equation (4) has been referred to as the slowness law and the Dieterich-Ruina law (Figure 6). In this equation, state, and thus friction, evolve even for truly stationary contact with $V = 0$, which has been referred to as aging [40-44].

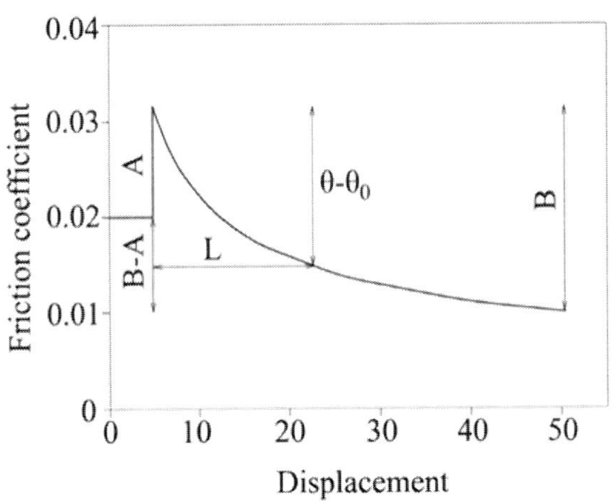

Figure 6. Dietierich-Ruina rate-and state-dependent friction law.

It should be emphasized that Dieterich-Ruina friction law is applicable not only for describing the contact of rock masses, but also for metals, paper and other materials. Nevertheless, despite its universality and wide applicability, some of the previous studies indicate that Dieterich-Ruina law could not be assumed for high sliding velocities, which could be explained by the occurrence of different mechanical and chemical reactions which arise due to frictional melting, which lubricates the contact surfaces. Such reactions include body heating, frictional melting and production of silicate gelatin or granular mass along the contact.

Except from Dietierich-Ruina friction law, there are other formulation of rock friction laws, that are also frequently assumed in the analysis of the spring-block model dynamics (For a very nice and thorough review on the laboratory-derived friction laws, please refer to Marone, [45]). For instance, Ruina [37] proposed a different evolution law in which velocity and displacement were of primary importance:

$$\mu(V,\theta) = \mu_0 + A\ln\left(\frac{V}{V_0} + B\ln\left(\frac{V_0\theta}{D_c}\right)\right)$$
$$\dot{\theta} = -\left(\frac{V\theta}{D_c}\right)\ln\left(\frac{V\theta}{D_c}\right) \tag{6}$$

A third law by Perrin, Rice, and Zheng [42] exhibits both aging and symmetry with respect to velocity changes:

$$\mu(V,\theta) = \mu_0 + A\ln\left(\frac{V}{V_0} + B\ln\left(\frac{V_0\theta}{2D_c}\right)\right)$$
$$\dot{\theta} = 1 - \left(\frac{V\theta}{2D_c}\right)^2 \tag{7}$$

Although the distinction between the two views of friction evolution is fundamental in terms of micromechanical interpretation of the underlying processes, the laws reproduce laboratory data in a similar fashion.

It has to be emphasized that the friction relations have been extended to include variations in normal stress [46-47], which we do not consider here.

For a mono-block model, Erickon et al. [34] revealed an occurrence of deterministic chaos for a certain parameter values, which is rather interesting result, concerning the dimensionality of the analyzed dynamical system. For a three-block model, chaotic solution occur for smaller values of control parameters when compared to the case of only a single block. Similar conclusion was drawn for the one-dimensional spring-block model in the continuum limit, including the presence of soliton and breather solutions [35].

As it can be seen from this brief review, phenomenological models are widely used in the study of fault motion. Except from these aforementioned models, there are also cellular automata models, two fractal overlap models or fiber bundle models (which represent a classical model of material failure rarely used in seismic analysis) and a dislocation model, which is still the most useful kinematic description of an earthquake.

The most commonly used cellular automata model as a representation of fault dynamics is suggested by Olami, Feder, and Christensen (1992). In general, cellular automata represent mathematical idealizations of physical system in which space and time are discrete, and it commonly consists of a regular uniform lattice, usually infinite in extent, with a discrete variable at each "cell". A cellular automaton evolves in discrete time steps, with the values of the variable at a single cell being affected by the values of variable at cells in its neighborhood on the previous time step. The variables at each site are updated simultaneously, based on the values of the variables in their neighborhood at the preceding time step, and according to a definite set of local rules. It is interesting to notice that cellular automata were originally introduced by von Neumann and Ulam (as a possible idealization of biological systems (1963), with the particular purpose of modeling biological self-reproduction.

Regarding the fractal models, a simple model was proposed by Chakrabarti and Stinchcombe (1999), in a form of a pair of dynamically overlapping Cantor sets, which allow one to study in detail the statistics of the overlap of one fractal object on another. The two fractal overlap magnitude changes in time as one fractal moves over the other. The overlap (magnitude) time series, which are changing as one fractal moves over the other, could be studied as a model time series of earthquake avalanche dynamics. In particular, the strain energy grown between the two surfaces due to a stick period is taken to be proportional to the overlap between them. Considering that slips occur at intervals proportional to the length corresponding to that area, previous studies indicated that a power-law, similar to Gutenberg-Richter law, for the frequency distribution of the energy release could be obtained.

As for the analysis of the effect of different perturbation on earthquake fault dynamics using phase response curves, such an approach has first been adopted in our recent paper [48], though the formalism related to phase description of nonlinear oscillators has already been invoked in the context of earthquake fault dynamics [24, 49-52]. So far, the phase response curves have often been used in theoretical neuroscience [53-56] and the general theory of coupled phase oscillators [57-60] as a tool to study the response of systems to external stimuli, as well as the ability of coupled systems to synchronize their activity.

Present chapter is majorly composed on the basis of the results of the following previous studies:

- Kostić, S., Franović, I., Todorović, K., Vasović, N. Nonlinear Dyn, 2013, 73, 1933-1943 [19]
- Vasović N., Kostić S., Franović I., Todorović K. Commun Nonlin Sci, 38, 117-129 [21]
- Kostić, S., Franović, I., Perc, M., Vasović, N., Todorović, K. Sci. Rep.: Nature Publishing Group, 2014, 4, 5401 [20]
- Kostić, S., Vasović, N., Franović, I., Todorović, K. Nonlinear Proc Geophys, 2013, 20, 857-865 [61]
- Kostić, S., Vasović, N., Franović, I., Todorović, K., Klinshov, V., Nekorkin, V. Submitted to Nonlinear Dynamics (under review), 2016 [22]
- Franović, I., Kostić S., Perc, M., Klinshov, V., Nekorkin, V., Kurths, J. Chaos, 2016, 26, 063105-1-12 [48]
- Kostić, S., Vasović, N., Franović, I., Todorović, K.. J Comput Nonlin Dyn 2013, 9, 031019 [62].

In present chapter, analysis of dynamics of spring-block models is conducted under the assumption that a single movement of a block or average displacement of a chain of blocks mimic the earthquake, which enables a qualitative analogy between the long-term dynamics of a block or a group of blocks and different regimes of fault motion. Thereby, transition between these regimes implies the occurrence of bifurcation due to subtle variation of control parameters, or under the impact of newly introduced factors. Considering the fact that sequence of earthquakes along a single fault is irregular in space, time and in size, a dynamics of a corresponding spring-block model should also be irregular. In case of a one-dimensional chain of blocks (i.e., interconnected blocks are moving in a single direction), one could observe the irregular dynamics only in time and size, which could correspond to the onset of deterministic chaos [35], or it could arise as a consequence of the effect of seismic noise [21].

3. NONLINEAR DYNAMICS – THEORETICAL BACKGROUND

In present chapter, nonlinear dynamics refers to the branch of physics that provides the description of complex natural phenomena using nonlinear differential equations. Moreover, we assume that such systems (dynamical systems) could be sensitive to initial conditions, which could further indicate a possible existence of a deterministic chaos. Although there is no formal and widely accepted definition of deterministic chaos, in present chapter we assume that deterministic chaos denotes the long-term aperiodic behavior of a dynamical system, which is highly sensitive to initial conditions (i.e. a small change of initial conditions commonly induce a large change in behavior of the observed dynamical system after a certain amount of time), and which, finite amount of time, converges towards the strange attractor. Famous mathematician, Ian Stewart, also defined it as the ability of simple models, without inbuilt random features to generate highly irregular behavior. In order for deterministic chaos to occur, observed dynamical system needs to go through a series of dynamical changes, i.e. bifurcations. There are three common ways how one could induce an onset of deterministic chaos: period-doubling, quasiperiodic and intermittent route to chaos. Period-doubling bifurcations occur at logistic mapping, where stable fixed point becomes unstable after a series of flip bifurcations (where fixed point loses its stability, with the occurrence of a limit cycle). Quasiperiodicity occurs for Van der Pol oscillator, where spiral node becomes unstable and a limit cycle occurs through a Hopf bifurcation. Torus appears after the second Hopf bifurcation. Further parameter change leads to decomposition of torus and the observed dynamical system settles on a constrained strange attractor, implying the occurrence of deterministic chaos. Such quasiperiodic route to chaos is also known as the Ruelle-Takens-Newhous route to chaos. The third, intermittent route to chaos is characterized by a saddle-node bifurcation in high-dimensional systems, where saddle node and a stable node coalesce and disappear, finally giving rise to trajectory with chaotic periods which are interspersed with regular oscillations.

It should be emphasized that besides differential equations, which describes the system's behavior in continuous time, there are also difference equations (iterative mapping), which describe the behavior of dynamical systems in discrete time intervals. Expect from this division of dynamical systems, there are also autonomous and nonautonomous equations, which describe explicit and implicit temporal dependence of the observed dynamical system, respectively. In current chapter, analyzed dynamical systems are described using autonomous ordinary first-order differential equations.

Dynamics of the systems observed in present chapter is usually determined by several control parameters, whose fine tuning induce specific transitions between different dynamical regimes, i.e. occurrence of bifurcations. Bifurcations represent qualitative, topological changes of the behavior of the observed dynamical system due to variation of values of control parameters. Number of parameters, whose tuning leads to the onset of bifurcation, commonly denotes the co-dimension of bifurcation. The most common bifurcations of co-dimension 1 are saddle-node bifurcation, transcritical and pitchfork bifurcation, Hopf bifurcation and period-doubling bifurcation. Homoclinic and heteroclinic bifurcations are the most frequent representatives of bifurcations of co-dimension 2. From the strict mathematical viewpoint, bifurcation occurs when eigenvalues of the characteristic equation of the observed system change their sign. In case characteristic equation has conjugate-complex solutions, real part of the solution needs to change the sign in order to induce a bifurcation. In present chapter, only the occurrence of a Hopf bifurcation is analyzed. This type of bifurcation actually represents a local bifurcation when a fixed point loses its stability and a pair of conjugate-complex eigenvalues of the linearized system in the neighborhood of the fixed point crosses the imaginary axis of the complex plane. Supercritical Hopf bifurcation occurs when the fixed point loses its stability with the parameter change, giving rise to the onset of a stable limit cycle. On the other hand, subcritical Hopf bifurcation arises when the increase of the value of control parameter further induce the coalescence of the stable fixed point and a stable limit cycle, eventually giving rise to the large stable limit cycle.

One should note that in present chapter only the results of the local bifurcation analysis are discussed. Global bifurcation, when larger dynamical sets, like periodic orbits, coalesce with the fixed point, or between each other, are analyzed only numerically.

There are two ways how models presented in current chapter are analyzed. One approach assumes linearization of the observed ordinary differential equations around the equilibrium state. Another approach, which is applied for stochastic differential equations, assumes the introduction of mean-field method.

Linearization of the observed dynamical systems (i.e. ordinary differential equations) is performed since according to Hartman-Grobman theorem, behavior of a dynamical system near the hyperbolic fixed point is qualitatively the same as the behavior of linearized system near the fixed point, under the condition that all eigenvalues of linearized system have the real part $Re(\lambda)\neq 0$. In that way, analysis of the behavior of dynamical system in a small neighborhood of the equilibrium state (fixed point) is more convenient after the performed linearization. After the performed linearization, one proceeds with determination of the characteristic equation of the observed system, which is further solved for eigenvalues λ_i. After the eigenvalues are determined, one continues with solving the relations between the examined control parameters of the observed system. Once these relations are determined, bifurcation curves are constructed, which denote the transition between different dynamical regimes. The next step assumes calculation of derivatives along the obtained bifurcation curves, In particular, in specific chosen points on the curve, one calculates the first derivative of the eigenvalue for the chosen control parameter. The obtained positive derivative indicates the existence of a direct Hopf bifurcation, while negative derivative denotes the inverse Hopf bifurcation. Unfortunately, local bifurcation analysis does not enable the determination of the exact type of Hopf bifurcation (whether it is a supercritical or subcritical), concerning the fact that global bifurcation could not be captured solely by the means of local analysis.

It is necessary to emphasize that, before conducting a linearization of the observed system, one should introduce convenient dimensionless parameters, especially in the case when a large number of potentially important parameters exist, so that their grouping into small number would make the analysis easier.

Regarding the application of the mean-field approach, we started from the initial assumption that modeling complex systems which comprise multiple characteristic spatial and time-scales inevitably incorporates a coarse-graining procedure, which consists in singling out the components that show a well controllable behavior into a few selected degrees of freedom, whereas the variations within the embedding environment, as well as the fluctuations associated to processes that take place at smaller spatial and temporal scales are approximated by introducing the different types of noise. Apart from noise, modeling of macroscopic systems often requires one to explicitly take into account the effects of interaction delays, which may derive from finite velocities of signal propagation and/or a latency of response of system's units to input variations. The separate or combined effects of noise and coupling delays may substantially alter the "bare" dynamics, unattended by these two ingredients. Among other effects of their interplay, one may find complex multistable behavior and competition between the delay-driven and the noise-induced oscillation modes.

The described models are mathematically stated in terms of systems of nonlinear stochastic delay-differential equations (SDDE). The analysis of stability and bifurcations of such systems presents a highly nontrivial problem. From qualitative perspective, the stochastic component may give rise to stochastic bifurcations, the kind of phenomenon fundamentally distinct from what is found in deterministic systems. The stochastic bifurcations have so far been characterized phenomenologically, in a sense that a certain time-averaged quantity, such as the asymptotic probability distributions of the relevant variables or the associated power spectra, may undergo some qualitative change under variation of noise intensity. For instance, it has been shown that the stochastic Hopf bifurcation from a stochastically stable fixed point to a stochastically stable limit cycle is accompanied by the loss of Gaussian property for the asymptotic distributions of the appropriate variables. In parallel, the degree by which the distribution departs from the normal one with super criticality depends on the particular system at hand. At variance with bifurcations in deterministic systems, where one may clearly identify a critical value of the control parameter, the change of system's behavior in noise-induced transitions is gradual.

The largest obstacle for analytically treating SDDE systems lies in the fact that the standard Fokker-Planck formalism cannot provide useful results concerning the qualitative analysis of stochastic stability and stochastic bifurcations of systems of SDDEs. The use of such an approach is severely constrained by the non-Markovian character and the nonlinearity of the underlying equations. In particular, the non-Markovian character is reflected in the occurrence of the so-called "conditional drift term". Even for univariate systems, the latter may only be resolved by an approximation, considering either the limit of a small delay or the limit of large delay, which substantially exceeds the system's correlation time. Nevertheless, for the setup involving multiple units, even the delay-free case under the assumption of molecular chaos cannot be handled analytically, though efficient numerical methods are available.

The failure of Fokker-Planck formalism implies the necessity for considering approximate methods, which should satisfy two criteria. On one hand, a viable approximate method should allow us to analyze the stability and stochastic bifurcations of the large systems of nonlinear SDDEs, whereas on the other hand, the approximation should be self-contained, in a sense that the parameter domains where it applies or fails are mathematically tractable.

One of the most frequently used approximate methods is the mean-field approach, which consists in deriving a low-dimensional reduced set of effective deterministic equations which may qualitatively account for the stability and bifurcations of the original set of N coupled SDDEs. The ensuing mean-field model is given by a set of delay-differential equations, where the influence of noise is manifested only by the noise intensity. The gains from introducing a mean-field model can in general be cast as twofold. On one hand, if it captures the behavior of the exact system with sufficient accuracy, the mean-field model substantially reduces the computational time for numerical integration, which in case of the exact system grows as $\approx N^2$ with its size. The other advantage lies in the ability to translate the highly nontrivial problem of stochastic bifurcations displayed by the exact system into a study of the compact deterministic mean-field system, which is amenable to standard bifurcation analysis, whereby the noise intensity acts as an additional bifurcation parameter.

In order to build a mean-field model for a general system of stochastic differential equations, one may adopt two different approaches. One approach is to consider the time dependence of a hierarchy of probability densities formally derived from the appropriate Fokker-Planck equation, whereas the other approach consists in focusing on the evolution of cumulants, whereby the full density of states is factorized into a series of marginal densities. In presence of coupling delay, the latter alternative is clearly a preferred one, and it further allows for a number of convenient approximations to be introduced in a controlled fashion. Note that one is bound to make certain simplifying approximations due to nonlinearity of the original system. In particular, the nonlinearity causes the cumulants of a given order to become linked to the ones of the higher order, which apparently renders the underlying series unclosed. The way to resolve this consists in truncating the series by introducing a closure hypothesis. A typical closure hypothesis integrates the cumulant approach with the Gaussian approximation, according to which all the cumulants above second order are assumed to vanish. By this approach, starting from the exact system of kN SDDEs, where k denotes the number of local degrees of freedom, one ends up with a set of $k(k + 3)/2$ deterministic delay-differential equations describing the collective dynamics in terms of the means, as well as the appropriate variances and the covariances.

The described mean-field method has so far mainly been applied in the field of neuroscience, where the use of effective low-dimensional approximate models has a long history. In particular, the mean-field method has proven successful in qualitative analysis of stability of collective dynamics in assemblies of delay-coupled noisy excitable elements. Nonetheless, it has also been demonstrated to successfully capture the scenarios for the onset and the delay-induced suppression of the collective mode, showing that the bifurcations displayed by the approximate model can qualitatively account for the stochastic bifurcations displayed by the exact system. It has also been indicated that the mean-field model can provide accurate quantitative predictions, reflected in a close agreement between its oscillation period and the average oscillation period of the exact system. Apart from systems with all-to-all (global) connectivity, the method has also been extended to cases with more complex connection topology, including random networks and paradigmatic modular networks. Note that the intricacies regarding the approximations behind the described mean-field approach, called the Gaussian approximation and the quasi-independence approximation, have been analyzed in detail, not only in terms of precise formulations, but also with respect to parameter domains that warrant their validity. In this context, it has been reported that the dynamics of the mean-field model can indicate in a self-consistent fashion the parameter domains where the mean-field approximations break down.

In light of the application of nonlinear dynamics in seismology, two different approaches are currently in the focus. One approach concerns the analysis of dynamics of spring-block models, which are assumed to mimic the movement along strike-slip faults in the Earth's crust. These models could be constrained to dynamics of a single block, or several blocks are mutually connected in a single array or within a two-dimensional plane. Previous studies have shown that distribution of displacements of a single unit or group of units follow characteristic seismological laws: Gutenberg-Richter and Omori-Utsu law. Analysis of such models also enables determination of key factors that control the onset of instability, which is considered to be analogous to the start of seismic fault motion. Another approach concerns the analysis of recordings of real earthquakes, or the analysis of the temporal and spatial distribution of main events, using method of nonlinear time series analysis. Such application could provide useful data for prediction of the size and location of the next great event (earthquake) or aftershock, and could be also advantageous for engineering design.

It should be emphasized that presented approach could be also used for modeling of movement along a reactivated fault, mainly due to the change of stress-strain conditions under the impact of deep underground excavation. Previous recordings and studies imply that earthquakes generated due to reactivated movement along the existing faults are by far the most frequent seismic feature in large mining fields. Moreover, such seismic events could occur with the magnitude up to M6, and with the intensity of VI-VIII degrees according to Modified Mercalli scale. There are many examples of such seismic events, including those at Lubin copper field and Silesia coal field in Poland (M4.5 and M4.3, respectively), at coal mines in France (up to M2.5), at coal mines in Germany (up to M4), atcoal mines in USA (up to M3.5) and at lead, silver and zinc mines in USA (up to M3). The necessary triggering factor for the reactivation of fault movement could occur as a continuous signal, originating from the work of large mining machines, or as a transient signal, due to effect of blasting and drilling at the excavation front. Such external perturbations could cause a change of the stress-strain conditions near the existing inactive fault, which could further induce a movement along the normal, reverse and strike-slip fault. Besides the activated movement, such changes of stress-strain conditions could also cause specific features, including rockbursts, pillar bursts, outbursts and bumps.

One should note that approach presented in current chapter is restricted to dimensionless phenomenological models. In particular, one cannot use the actually measured values for the analysis, but examined models could serve only for qualitative assessment of dynamics of the earthquake nucleation process. Reason for this lies in the fact that analyzed spring-block models, for a certain parameter values, exhibit stick-slip motion, which resembles the seismogenic fault dynamics. Also, such models are very convenient for qualitative analysis of the effect of different factors, including background seismic noise, transient impact of a passing seismic wave from near or distant earthquake or the influence of different types of interaction between different parts of the fault.

Other techniques of nonlinear dynamics, like nonlinear time series analysis, could be used to examine the recorded accelerograms of the past earthquakes. Such time series actually represent the motion of the ground surface due to passage of seismic waves, and the results of such analysis could be directly beneficial for urban planning and civil engineering. This techniques include, but are not restricted to: false nearest neighbor method, mutual information method, deterministic test, stationarity test, surrogate data testing or calculation of maximal Lyapunov exponent. This techniques could also be used for the analysis of time series of different origin, including volcanic seismicity, blasting, etc. In general, application of these methods should lead to full characterization of recorded signal, which could further enable its prediction, at least for the near future.

3. Dynamics of a Mono-Block Model

Kostić et al. [19] analyzed model proposed by Madariaga [34], which describes the motion of a single block along the rough surface coupled with Dieterich- Ruina rate and state dependent friction law:

$$\dot{\theta} = -\left(\frac{v}{L}\right)\left(\theta + B\,log\left(\frac{v}{v_0}\right)\right)$$
$$\dot{u} = v - v_0 \qquad (8)$$
$$\dot{v} = \left(-\frac{1}{M}\right)\left(ku + \theta + A\,log\left(\frac{v}{v_0}\right)\right)$$

with the following parameter used:

- *k* - spring stiffness,
- M - mass of the block,
- L (Dc) - critical sliding distance necessary to replace the population of asperity contacts,
- A, B - empirical constants (dependent on the material properties).

For convenience, system (8) could be formulated in non-dimensional fashion by defining the new variables $\theta,'$ $v,'$ u' and t': $\theta = A\theta,'$ $v = v_0 v,'$ $u = Lu,'$ $t = (L/v_0)t.'$ After returning to the use of θ, v, u and t, system (8) becomes:

$$\dot{\theta} = -v\left(\theta + (1+\varepsilon)\,log(v)\right)$$
$$\dot{u} = v - 1 \qquad (9)$$
$$\dot{v} = -\gamma^2\left[u + (1/\xi)\left(\theta + log(v)\right)\right]$$

where $\varepsilon = (B-A)/A$ measures the sensitivity of the velocity relaxation, $\xi = (kL)/A$ is the nondimensional spring constant, and $\gamma = (k/M)^{1/2}(L/v_0)$ is the nondimensional frequency [34].

3.1. Frictional Memory Effect

Frictional memory effect is analyzed by introducing the time delay τ in system (9) [19]:

$$\begin{aligned} \dot{\theta} &= -\nu\left[(\theta+(1+\varepsilon)\log(\nu(t-\tau))\right] \\ \dot{u} &= \nu-1 \\ \dot{\nu} &= -\gamma^2\left[u+(1/\xi)(\theta+\log(\nu))\right] \end{aligned} \tag{10}$$

The system of first-order ordinary differential equations (10) has a unique stable stationary solution (θ, U, V) = (0, 0, 1), which corresponds to the steady sliding of the block. Hence, by this model, the fault is in a state of constant motion, which could correspond to some plate boundary segments, like an Eastern California shear zone, where the average repeat time of fault seismic motion is of the order of thousands of years [20].

The main motivation for including the time delay in friction term lies in the nature of the rock friction itself – change in velocity of the block does not instantaneously induce a corresponding change. On the contrary, there is a certain time delay between the change of velocity and change of friction, which could be modeled as it is suggested in Eq. (10). Another motivation for the introduction of time delay comes from the similarity of the dynamics of spring-block model with the relaxation oscillator. In particular, stick-slip motion represents an inherent property of the relaxation oscillators, whose main mark is the existence of two time scales: the slow one, during the energy accumulation (the so-called stick phase) and the fast one, during the energy release (to so-called slip phase).

Standard bifurcation analysis leads to the following expression for ε as a function of ω:

$$\varepsilon = -1 + \sqrt{\frac{\left[\omega^3 - \omega\gamma^2\left(\frac{1+\xi}{\xi}\right)\right]^2 + \left[\omega^2\left(\frac{\gamma^2}{\xi}+1\right) - \gamma^2\right]^2}{\left(\omega\frac{\gamma^2}{\xi}\right)^2}} \tag{11}$$

and for τ as a function of ω:

$$\tau = \tau_c = \frac{1}{\omega}\left[\operatorname{arctg}\left(\frac{-\omega^2\left(\frac{\gamma^2}{\xi}+1\right)+\gamma^2}{-\omega^3+\omega\gamma^2\left(\frac{1+\xi}{\xi}\right)}\right)+k\pi\right] \tag{12}$$

where k is any nonnegative integer such that $\tau_k \geq 0$.

The above parametric equations for ε, ξ and τ coincide with the Hopf bifurcation curves illustrated in Figure 7, where the bifurcation curves $\tau(\varepsilon)$ are shown at the fixed parameter values $\xi = 0.5$ and $\gamma = 0.8$. Two different scenarios could occur when increasing the value of time delay. On one hand, the increase of time delay could induce the so-called amplitude or oscillatory death, indicating a transition between the different regimes of periodic oscillations, or the return to equilibrium state, respectively. On the other hand, increase of frictional delay could destabilize the system, with the final occurrence of deterministic chaos. Apparently, the system exhibits quasiperiodic route to chaos [63-64], such that after two supercritical Hopf bifurcations, the dynamics falls onto the strange attractor. Apparently, only by increasing the time-lag τ, e.g., by setting $\tau = 0$, $\tau = 10$, $\tau = 12$, to $\tau = 20$, and by slightly changing the other parameter values, the block dynamics changes from the fixed point, over the limit cycle oscillation (first Hopf bifurcation) and torus (second Hopf bifurcation) to chaos (Figure 8).

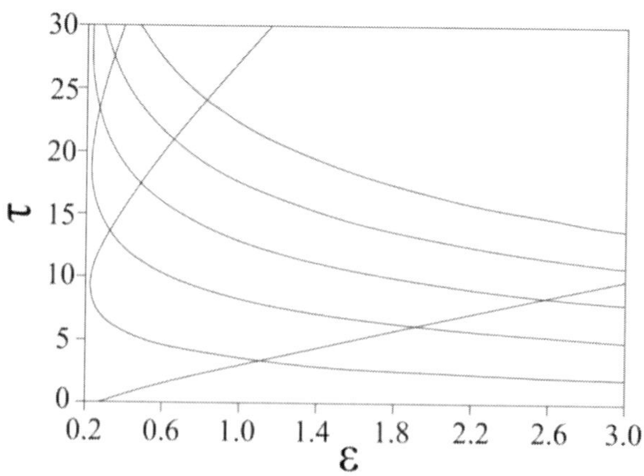

Figure 7. Hopf bifurcation curves $\tau(\varepsilon)$, for the fixed values of parameters $\xi = 0.48$, and $\gamma = 0.8$. The appropriate time series and the phase diagrams corresponding to particular points are shown in Figure 8.

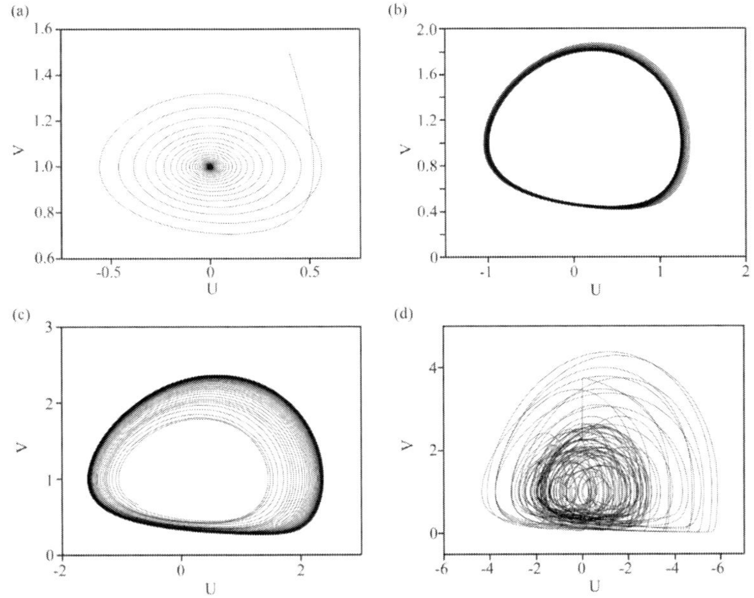

Figure 8. Phase portraits of variable v for a. $\tau = 0$, $\varepsilon = 0.23$, $\xi = 0.48$ and $\gamma = 0.8$ (fixed point); b. $\tau = 10$, $\varepsilon = 0.3$, $\xi = 0.48$ and $\gamma = 0.8$ (limit cycle); c. $\tau = 12$, $\varepsilon = 0.5$, $\xi = 0.48$ and $\gamma = 0.8$ (torus); d. $\tau = 20$, $\varepsilon = 0.5$, $\xi = 0.48$ and $\gamma = 0.8$ (deterministic chaos). Parameter values are determined by the position of Hopf bifurcation curves in Figure 7.

3.2. Impact of Transient Perturbations

Kostić et al. [20] also examined the effect of a passing seismic wave on the dynamics of a spring-block model by introducing the following transient perturbation of the block acceleration:

$$\begin{aligned}
\dot{\theta} &= -V(\theta + (1+\varepsilon)log(V)) \\
\dot{U} &= V - 1 \\
\dot{V} &= -\gamma^2[U + (1/\xi)(\theta + log(V))] + A sin\left(\frac{2\pi t}{T}\right)e^{\{[(t-t_0)/t_w]^n\}}
\end{aligned}, \qquad (13)$$

with the following parameters:

- A - maximum amplitude of the induced change,

- T - wave period,
- t_0 - moment for which the perturbation amplitude reaches its maximum value,
- t_w - half width of the induced perturbation,
- n - integer that controls the rise-time.

In terms of real earthquakes, more rapidly varying (short-period) transient changes of block acceleration could correspond to the effect of closer, smaller earthquakes, while more slowly varying (long-period) transient changes might describe larger, more remote earthquakes [20]. Parameters are set with respect to the periodic oscillations of the unperturbed system (13) just above the bifurcation curve ($A_0 = 0.6$, $T_0 \approx 12$). In this context, we have considered the short-period ($T < 12$) and long-period ($T > 12$) acceleration changes of different duration $t_w \in [3, 100]$. Kostić et al. [20] conducted the numerical analysis of the system (12) using Runge-Kutta fourth order system by assuming parameter values before and after the bifurcation curve. In the former case, the only relevant result occur when introducing the long-term transient changes which cause weakly irregular oscillations, that could be treated as a example of transient chaotic behavior, indicating the mechanism of instantaneous earthquake triggering. In the latter case, long-term high-frequent transient acceleration change induces the onset of quasiperiodic-like motion with modulated amplitude, indicating a possible transition to a precursory creep regime. On the other hand, longer low-frequent transient perturbations induce irregular oscillations which could correspond to transient seismic motion along the fault.

The main starting assumption of the work by Kostić et al. [20] is that different stages of block motion separated by the Hopf bifurcation curve correspond to the different creeping regimes along the fault. In particular, the uniform motion of the block is treated as a model of fault slip in the post-seismic creep regime, just after the last seismic event has occurred. The periodic oscillatory motion is considered to correspond to inter-seismic creep regime, which could be characteristic of more "mature" faults during the period of slow accumulation of elastic. One should note that model (13) only describes the different stages of the creeping aseismic displacement along the fault, whereas the onset of transient quasiperiodic-like (precursory creep regime) and irregular chaos-like motion (coseismic dynamics) occurs only under the effect of external perturbation. The two latter types of motion are assumed to correspond to the precursory creep regime and the seismic motion, respectively. In this context, Kostić et al. [20] analyzed only the possibility of generating the complex behavior, while common approach differs assumes that the dynamic instability arises when the sliding velocity exceeds some pre-defined threshold value [65-69].

Under the assumption that a passing seismic wave propagates along the fault, dynamics of system (12) reveals the following picture. In case when the generated change is of longer duration (t_w = 100), high-frequent acceleration changes in the inter-seismic creep-regime lead to quasiperiodic-like dynamics (precursory creep regime), while long-period changes of acceleration in both post-seismic and inter-seismic creep regimes invoke irregular oscillations or chaos-like behavior (seismic motion). It is clear that the longer low-frequent transients could excite a seismic motion even if the fault is in an uniform creep regime. Such type of triggering has been observed in real conditions [70-71].

Significantly longer acceleration changes (e.g., t_w = 800) induce a temporary transition from oscillatory to uniform motion, after which the system converges to a stable limit cycle. San Andreas fault system, for example, did not experience significant triggering by the Landers earthquake, since its thicker and more deformed fault gouge requires greater stress change to induce a rupture [72].

The performed analysis has confirmed the occurrence of immediate or instantaneous dynamic triggering effect of a passing seismic wave, which has already been observed in situ following Landers [73], Denali [74] and Izmit earthquake [70]. Kostić et al. [20] determined that the long-period transients, which are assumed to correspond to the effect of distant earthquakes, have the strongest impact on the fault motion, leading to irregular chaos-like behaviors. On the other hand, the high-frequent transients, presumably originating from the nearby earthquakes, induce a transition to quasiperiodic-like motion (precursory creep regime).

It should be emphasized that the analysis performed by Kostić et al. [20] revealed that the low-amplitude variations have no significant impact on the dynamics of system (14). Nevertheless, previous studies indicated that even very small dynamic changes (down to $3x10^{-9}$) could trigger faults that are sufficiently near failure [75-76]. This discrepancy could be explained by the fact that model analyzed by Kostić et al. [20] represents only the fault creeping that is relatively stable with respect to the short-lived perturbations and that the dynamics of this model does not exhibit transient or asymptotic instability without the external perturbation of sufficiently high amplitude.

The results of Kostić et al. [20] are consistent with the work of Gomberg et al. [77-78], who indicated that the transient dynamic stresses could not explain the delay triggering. Also, Belardinelli et al. [79] found that a dynamic stress pulse is able to promote a nearly instantaneous failure, but cannot induce delayed triggering.

3.3. Impact of Continuous Perturbations

In the work of Kostić et al. [62], effect of continuous perturbations is analyzed by assuming the time dependent character of ε and ξ in the following way:

$$\begin{aligned}
\dot{\theta} &= -V\left[\theta + \left(1 + \varepsilon + \delta_{\varepsilon}\sin(\omega_{\varepsilon} t)\right)\log V\right] \\
\dot{U} &= V - 1 \\
\dot{V} &= -\gamma^2\left[U + \left(1/\xi + \delta_{\xi}\sin(\omega_{\xi} t)\right)(\theta + \log V)\right]
\end{aligned} \tag{14}$$

where δ_ε, δ_ξ, ω_ε, and ω_ξ represent the constant oscillation amplitudes and the angular frequencies, respectively ($\delta_\varepsilon \leq \varepsilon$, $\delta_\xi \leq \xi$). It is assumed that external perturbations induce small oscillations near the equilibrium values of some of the system parameters, but admitting the periodic motion, which could arise from slow rupture along the faults or to some non-natural source of vibrations, on one side [80], or Earth tides and reservoir effects, on the other side [81].

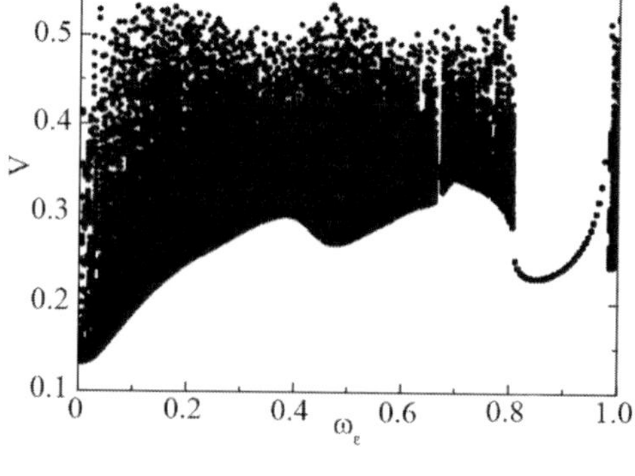

Figure 9. Projection of the attractors in the Poincare section onto the V-coordinate vs the bifurcation parameter ω_ε. The sections are provided for the plane $\theta = 1$ and ω_ε and with the sampling step of 0.01, obtained after $8\mathrm{x}10^6$ iteration steps for the fixed parameters admitting the limit cycle: $\varepsilon = 0.4$ ($\delta_\varepsilon = 0.3$), $\xi = 0.5$ and $\gamma = 0.7$.

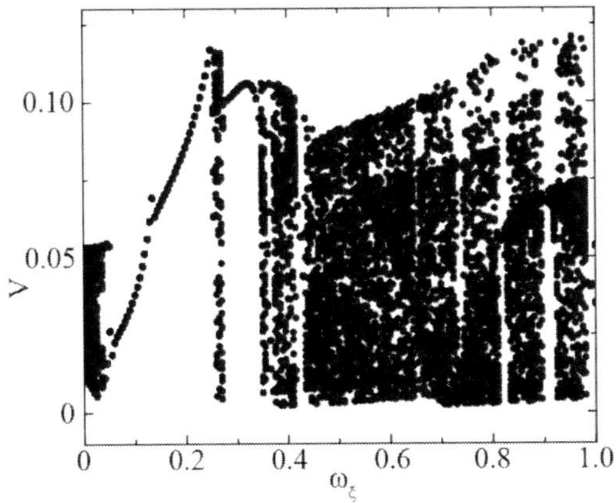

Figure 10. V-coordinates of the respective Poincare sections under the variation of ω_ξ. The sections are provided for the plane $\theta = 1$, having sampled ω_ξ with the 0.01 step size, obtained after $4x10^6$ time steps, for the fixed parameter values admitting the limit cycle: $\varepsilon = 0.4$, $\xi = 0.1$ ($\delta_\xi = 0.3$) and $\gamma = 0.7$.

If only a single parameter, ε or ξ, undergoes periodic oscillations, while the other parameter is fixed, it turns out that as the parameter ω_ε takes smaller values at each step in the range [0, 1], a transition from periodic motion to deterministic chaos is observed, with periodic and quasiperiodic windows interspersed between chaotic clouds of dots (Figure 9).

The analysis of the system's dynamics when the perturbation of the parameter ξ (ω_ξ) is assumed (also for the fixed upper value of amplitude, $\delta_\xi = 0.4$) indicates more complex picture, with successive transition from periodic and quasiperiodic motion to deterministic chaos, interspersed with periodic and quasiperiodic windows (Figure 10).

Although the onset of chaos is observed due to a single parameter perturbation, the drawback of such an approach is that the associated perturbation amplitude has to be comparable to the parameter's equilibrium value ($\delta_\varepsilon = \delta_\xi = 0.4$). This is why we conducted additional analysis of complex dynamics of the examined model by choosing a value of angular frequency admitting the periodic motion of the block in an unperturbed state, while the dynamics is observed by changing the perturbation amplitudes for each of the observed parameter separately (Figure 11).

As it is clear from Figure 11 (upper figures), in case of perturbing only a single parameter (ε or ξ), there is a direct transition to deterministic chaos for both amplitude values, δ_ε and δ_ξ, in the range [0, 0.4]. This type of scenario to chaos could imply the existence of some global bifurcation.

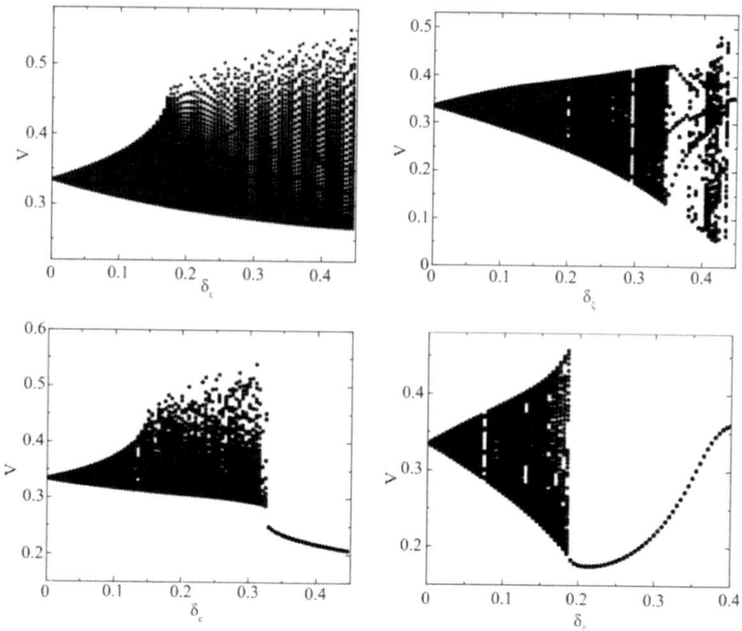

Figure 11. Orbital diagrams for the system (14) under the variation of the perturbation amplitude δ_ε and δ_ξ, with the associated frequencies fixed at $\omega_\varepsilon = \omega_\xi = 0.3$ (upper figures) and $\omega_\varepsilon = \omega_\xi = 0.8$ (lower figures), constructed for the plane $\theta = 1$, adopting the 0.01 sampling step sfter $8x10^6$ iteration steps. The remaining parameters admit the limit cycle: $\varepsilon = 0.4$, $\xi = 0.5$ and $\gamma = 0.7$.

In the work of Kostić et al. [62] they also analyzed a case of a single amplitude perturbation, only this time with higher frequency values ($\omega_\varepsilon = \omega_\xi = 0.9$). As it can be seen in Figure 11 (lower figures), for the higher frequency values, a transition from periodic motion to deterministic chaos is observed. By comparing orbital diagrams in Figure 11 (lower figures), it is clear that the onset of chaos could be controlled only by tuning the angular frequencies of assumed periodic perturbation of the selected parameters (ε or ξ).

One should note that although the assumed perturbations are rather simple, the performed analysis showed that the onset of deterministic chaos could be observed for smaller values of the control parameter ε, in comparison to the results of Erickson et al. [34], who observed irregular aperiodic oscillations for $\varepsilon = 11$. On the other side, results of Kostić et a. [62] are comparable with the research also conducted by Erickson et al. [35], where the transition to chaos is observed for $\varepsilon = 0.5$.

4. Dynamics of a Chain of Blocks

4.1. Global Coupling

Vasović et al. [21] examined the chain of all-to-all coupled blocks (each of i = 1, 2,…, N units is coupled with each other unit):

$$\dot{U}_{1i}(t) = U_{2i}(t)$$

$$dU_{2i}(t) = \left\{ -U_{1i}(t) + \Phi(U_{2i} + v) - \Phi(v) + \frac{K}{N}\sum_{j=1}^{N}\left[U_{1j}(t-\tau) - U_{1i}(t)\right] \right\} dt + \sqrt{2D}dW_i \, , \tag{15}$$

where:

- U_{1i}, U_{2i} - displacement and velocity of the i-th block, respectively,
- K – spring constant,
- Φ - friction force, $\Phi(V) = -(\mu_0 + a\ln(V))$ (μ_0 - steady-state friction, a - material property),
- τ - time delay,
- v - nondimensional pulling background velocity,
- $\sqrt{2D}dW_i$ - stochastic increments of independent Wiener process, $E(dW_i) = 0, E(dW_i dW_j) = \delta_{ij}dt$, ($E()$ - expectation over many realizations of the stochastic process, D - intensity of additive local noise).

One should note that rate-dependent nature of friction law has been indicate by previous studies [82-83].

Starting from model (15), Vasović et al. [21] developed the following averaged mean-field model:

$$
\begin{aligned}
&\dot{m}_{U_1}(t) = m_{U_2}(t),\\
&\dot{m}_{U_2}(t) = -m_{U_1}(t) + a\ln\nu - a\ln\left(m_{U_2}+\nu\right) + \frac{a}{2}\frac{1}{\left(m_{U_2}+\nu\right)^2}s_{U_2} + \frac{a}{4}\frac{1}{\left(m_{U_2}+\nu\right)^4}3s_{U_2}^2 +\\
&+ K\left[m_{U_1}(t-\tau) - m_{U_1}(t)\right]\\
&\frac{1}{2}\dot{s}_{U_1}(t) = s_{U_1U_2}\\
&\frac{1}{2}\dot{s}_{U_2}(t) = s_{U_2}\left[-\frac{a}{\left(m_{U_2}+\nu\right)} - \frac{a}{\left(m_{U_2}+\nu\right)^3}s_{U_2}\right] - (K+1)s_{U_1U_2} + D,\\
&\dot{s}_{U_1U_2} = -s_{U_1U_2}\left[\frac{a}{\left(m_{U_2}+\nu\right)} + \frac{a}{\left(m_{U_2}+\nu\right)^3}s_{U_2}\right] - (K+1)s_{U_1} + s_{U_2}(t)
\end{aligned}
\tag{16}
$$

where: $m_{U_1}(t)=\langle U_1(t)\rangle, m_{U_2}(t)=\langle U_2(t)\rangle, s_{U_1}(t)=\langle n_{U_1}^2(t)\rangle, s_{U_2}(t)=\langle n_{U_2}^2(t)\rangle, s_{U_1U_2}(t)=\langle n_{U_2}\cdot n_{U_1}\rangle$, and $n_{U_j}(t)=m_{U_j}(t)-U_{ji}(t)$, j = 1,2. In their work, Vasović et al. [21] showed mean-field approximated model (16) describes the dynamics of the starting model (15) accurately enough, so the obtained bifurcation curves in general correspond to bifurcations of the original system. Local bifurcation analysis of system (15) indicates a transition from equilibrium state to periodic oscillations for certain parameter values (Figure 12). From the seismological viewpoint, fixed point corresponds to absence of movement or steady sliding along the fault, while limit cycle denotes the aseismic creep along the fault [84-85].

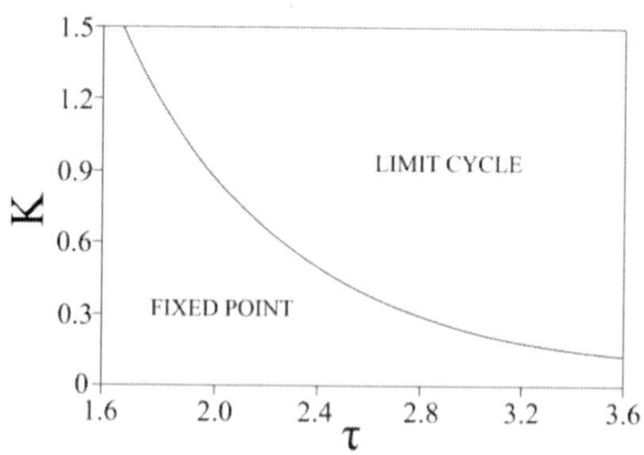

Figure 12. Parameter domain (K,τ) admitting fixed point or limit cycle of the mean-field approximated model (16). For a given parameter domain, other parameters are held constant: ν = 1.2, D = 0.002, a = 0.1.

According to Vasović et al. [21], derived model (16) is in equilibrium state when the initial conditions are set near the fixed point. However, when initial conditions are assumed to take values away from the equilibrium point, mean-field model (16) could be both in equilibrium state or exhibit periodic regular oscillations, as a consequence of a global bifurcation or due to impact of the introduced random seismic noise when the starting stochastic system is near the bifurcation curve.

From the viewpoint of seismology, performed bifurcation analysis indicates that earthquakes could be expected to occur only in a bi-stable dynamical regime in the vicinity of a bifurcation curve provided that initial conditions along the fault are far from the equilibrium state (the case of active fault). In particular, irregular oscillations with amplitude higher than assumed noise level arise only for this regime, as an effect of either the introduced noise or the global attractor. Regular periodic oscillations of the starting system (16) above the bifurcation curve rarely occur in real conditions, since periodic occurrence of earthquakes is captured in the real conditions only for the strongest earthquakes at certain locations in the Earth's crust [86]. However, occurrence of all earthquakes within certain area is aperiodic. This means that periodic oscillations also represent an example of aseismic creep, which is frequently observed in real conditions [84-85].

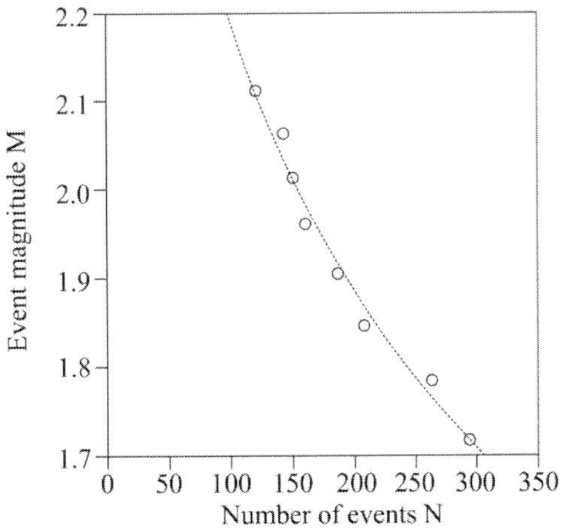

Figure 13. Examples of power-law behavior of system (15) away from the equilibrium point in a bi-stable regime near the bifurcation curve. It is clear that for different parameter values magnitude-frequency distribution follows the Gutenberg-Richter law ($b = 0.99$) with satisfying statistical accuracy ($a = 0{,}1$; $D = 0.001$; $\tau = 2$; $K = 0.865$; $\nu = 1.2$).

According to Vasović et al. [21], performed statistical analysis indicated that distribution of observed event magnitudes follows the Gutenberg-Richter power law with high values of R^2 and low residual sum of squares and with the b-value $\approx$ 0.99 (Figure 13). Similar power law behavior is observed in neural networks [87]. In this case, event magnitude is defined as a natural logarithm of displacement sums for all blocks.

4.2. Varying Neighboring Interactions

Kostić et al. [22] analyzed a system comprised of N blocks, whereby each block i interacts with 2K nearest neighbors:

$$\dot{u}_i = v_i$$

$$\dot{v}_i = -u_i + \Phi\left(v_i + v_0\right) - \Phi(v_0) + \frac{C}{N}\sum_{j \in J}\left(u_{i+j}(t-\tau) - u_i\right) + \sqrt{2D}\xi_i(t), \tag{17}$$

where J denotes the set of indices $J = \{-K,...,K\} \setminus \{0\}$, C is the coupling strength, τ is uniform interaction delay, often previously assumed in the case of neural chains [88-90], v_0 is the pulling velocity of the upper moving plate, $\xi_i(t)$ - independent Gaussian white noise ($\langle \xi_i \rangle = 0, \langle \xi_i \xi_j \rangle = \delta_{ij}$), deriving from various sources [91-92] and Φ is the frictional force $\Phi(v) = \mu_0 - a\ln(v)$, where μ_0 and a are the same as in Equation (15).

Impact of coupling strength on the dynamics of the observed system arises from the effect of the crushed material along the fault zone, which commonly represents a cohesionless unconsolidated material composed of finely crushed particles. Influence of this material primarily depends on its composition, zone width and fault depth.

One should note that blocks at the end of the array are connected to neighboring blocks only on one side.

Starting from model (16), Kostić et al. [22] derive the following mean-field approximated model with qualitatively the same dynamics as in Eq. (17):

$$\dot{m}_u(t) = m_v(t)$$
$$\dot{m}_v(t) = -m_u(t) + a\,ln(\nu_0) - a\,ln(m_v + \nu_0) + \frac{1}{2}\frac{D}{(m_v + \nu_0)} + \frac{2KC}{N}(m_u(t-\tau) - m_u(t)) \tag{18}$$

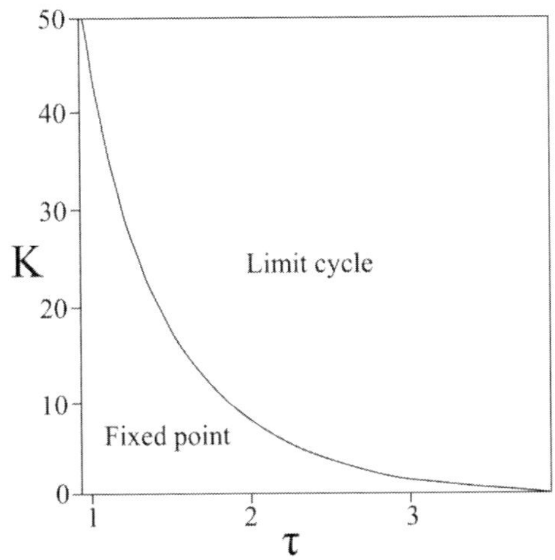

Figure 14. *K(τ)* bifurcation curve describing the destabilization of fixed point in the mean-field model (18). The remaining system parameters are fixed at C = 5, D = 0.0001, a = 0.1, v0 = 1.2.

Analysis performed by Kostić et al. [22] showed that a supercritical Hopf bifurcation occurs indicating a transition from fixed point to a limit cycle, solely by increasing the time delay in interaction (Figure 14). For relatively high values of τ, bifurcation can arise even in case of very small range of interactions (K = 3). On the other hand, bifurcation could occur only by increasing the number of interacting blocks for a constant τ. From the seismological viewpoint, this means that seismic fault movement could be induced solely by increasing the active length under the condition that there is a delayed interaction among different fault segments [22].

According to Kostić et al. [22], effect of seismic noise on dynamics of system (17) is negligible for the common ratio of coseismic slip rate vs. seismic noise is in the range 10^{-2}-10^{-7}.

One should note that bifurcation in the starting stochastic system (17) occurs solely by increasing the coupling strength C provided the delayed interaction between the units.

Furthermore, the results of analysis performed by Kostić et al. [22] obtained indicate the occurrence of an inverse supercritical Hopf bifurcation, induced by the decrease the friction, which is expected in real conditions, since friction usually decreases with the increase of fault motion velocity. This means that a reduction of friction along the fault zone under the impact of pore water or similar, could induce the onset of bifurcation, under the assumed delayed interactions among different fault segments.

According to Kostić et al. [22], a group of coupled blocks with the largest displacements in the considered array (17) could be considered as a model for a hypocenter of a seismic event. In this case, they assume that event magnitude is defined as a natural logarithm of a sum of squares of displacements:

$$M = \ln\left(\sum_{i=1}^{N}(u_i)^2\right) \tag{19}$$

where N is the total number of blocks in the observed system. Here, Kostić et al. [22] take the sum of squared displacement for each block as a measure of accumulated potential energy released during a single event. One should note that this is in no way in contrast to the assumptions made regarding the Figure 13. In essence, distribution of events remains the same. Nevertheless, authors consider the equation (18) to be a more reasonable assumption, concerning the definition of the potential energy itself. The distribution of the events with the largest magnitudes of displacements, is nearly periodic, so the suggested model (17) could be treated as a periodic or characteristic earthquake model [93].

5. Phase Response Curves for Earthquake Fault Dynamics

The fault dynamics is often represented by the class of spring-slider block models incorporating different forms of constitutive friction laws. Such models can qualitatively account for the relevant regimes of fault dynamics, including the aseismic creep motion or the stick-slip motion, whereby the latter is signature for earthquakes. The studies based on these models have so far mainly aspired to gain insight into the scaling laws and the occurrence of characteristic events, as well as to elucidate the relation between small and large earthquakes. A substantially different approach is adopted in the work of Franović et al. [48]. Rather than examining the statistical features of the underlying time series or assessing the earthquake hazard, they to analyze a representative class of fault models from the perspective of nonlinear dynamics theory.

The physical background for taking such an approach can be described as follows. Apart from the scale-invariant statistical features reflected in several well-known empirical scaling laws, certain earthquakes may exhibit characteristic features, manifested in "characteristic earthquakes" with the well-defined time or energy scales [24, 93-94]. The presence of a typical timescale implies that the fault fails in a pseudo-periodic series, making its dynamics reminiscent of that of a relaxation oscillator. The underlying oscillations are likely to be sensitive to external perturbations, which may derive from various forms of additional forcing whose duration and magnitude are small compared to the tectonic load. In general, if the perturbation is sufficiently small, it does not influence the amplitude of oscillations, but may considerably affect their phase. Then, an important conceptual issue, especially from the seismological point of view, becomes to determine whether and how the phase of stick-slip oscillations, and thereby the characteristic event itself, is delayed or advanced by such perturbations.

The motion of blocks comprising the complex fault is given by a simple generalization of the form:

$$\begin{aligned}
\frac{d\theta_i}{dt} &= -v_i\left(\theta_i + (1+\varepsilon_i) ln\, v_i\right) \\
\frac{du_i}{dt} &= v_i - 1 \\
\frac{dv_i}{dt} &= \gamma_i^2\left(c(u_i - u_j) + u_i + (1/\xi_i)(\theta_i + ln\, v_i)\right)
\end{aligned} \tag{20}$$

where $i, j \in \{1, 2\}$, $i \neq j$ while c denotes the coupling strength.

As already indicated, the pseudo-periodic recurrence times have primarily been associated with large characteristic earthquakes [52, 95-99]. By one scenario, the latter involve breaking of the most part of or the entire seismogenic zone [24, 93, 100-101], whereas by the other scenario they emerge due to breaking of similar sections of complex faults [24, 102]. Well-known examples are earthquakes in the Nankaido region (Japan), the northern, southern and Parkfield sections of the San Andreas Fault [103], and several regions in China [104-105]. Apart from these large characteristic earthquakes, the description of fault dynamics in terms of relaxation oscillator models may also be justified for certain small repeating earthquakes [106]. One should note that many of the relevant models of fault dynamics may yield periodic sequences of events or series with a strong periodic component. For instance, such behavior has been found for the one- and two-dimensional versions of the Burridge-Knopoff model [107-109], as well as in case of the Olami-Feder-Christensen model [110]. It has also been suggested that models of coupled relaxation oscillators displaying the stick-slip dynamics may account for a phase-locking mechanism behind earthquake clusters. The latter conform to rupture patterns where the main events occur in groups comprising nearby or distributed faults with similar characteristic periods [49-50, 111].

5.1. Phase Response Curves for the Monoblock Fault

5.1.1. Theoretical Background

Phase response curve is an inherent feature of an arbitrary oscillator, which reflects its sensitivity to a brief pulse-like stimulus. Phase response curve is given by the dependence of the phase shift, induced by a perturbation, as a function of the oscillation phase at which the perturbation has occurred. The effect of phase resetting due to pulse perturbation may formally be treated as follows. Let us first consider a one-dimensional oscillator, described only by a continuously increasing phase variable φ which evolves as $\dot{\varphi}=\omega$. Then the system's phase just after a pulse stimulus of strength κ introduced at the moment t_p can be written as [18, 55].

$$\varphi_+(t_p)=\varphi(t_p)+\kappa Z(\varphi,\kappa) \qquad (21)$$

where $Z(\varphi,\kappa)$ stands for the phase response cruve. In a more general case, periodic oscillations are characterized by a limit-cycle attractor $X_0(t)$ in an N-dimensional phase space. Nevertheless, the notion of isochrones [112-113] still allows one to consider a phase space representation of the form (a,φ) where a is an (N-1) - dimensional "amplitude," and φ is the regular phase variable obeying $\dot{\varphi}=\omega$ [55]. Without loss of generality, one may assume that the "amplitude" vanishes on the limit cycle (a = 0). In this setup, if a kick is introduced at the moment t_p, the reset of the state (a, φ) just after t_p may be expressed as [55]:

$$\mathbf{a}_+(tp)=\mathbf{a}(t_p)+\kappa A(\mathbf{a}(t_p),\varphi(t_p),\kappa)=\kappa A(0,\varphi(t_p),\kappa)$$
$$\varphi_+(t_p)=\varphi(t_p)+\kappa\Phi(\mathbf{a}(t_p),\varphi(t_p),\kappa)=\varphi(t_p)+\kappa Z(\varphi(t_p),\kappa) \tag{22}$$

The above equations take into account that the initial state lies on the limit cycle (a = 0), such that $\Phi(0, \varphi, \kappa) = Z(\varphi, \kappa)$ holds. In terms of application, phase response curves were first introduced in the study of oscillations in biological systems, including cardiac cells, fireflies populations and especially neural networks [114-120]. Within these fields, as well the general theory of coupled phase oscillators, the method has facilitated an analysis of the units' interaction properties, including stability, synchronization or clustering. The concept of phase response curves allows one to reduce complex models of oscillators to simple phase models which still reflect important features of the original oscillators, such as the point that the effect of perturbation depends on the dynamical state of the oscillator. By this physical picture, each oscillator possesses a characteristic phase response curve that can be computed numerically or measured experimentally [121-123].

In a general multidimensional system, the kick may be applied to any of the system variables. In the work of Franović et al. [48], a perturbation is added to the second equation of the system (20), which is the most plausible choice, because it may be interpreted as a small variation at the loading point. The corresponding equation then takes the form $du/dt = v\text{-}1+f(t)$, where $f(t)$ is the perturbation term. In real faults, such perturbations may derive from various natural and artificial sources, including rock break, pressure fluctuations or crack vibration due to movement of magma and volcanic gases [124], sudden stress drops [125-126] drilling and blasting in underground mining activities [127-128], as well as micro earthquakes due to hydraulic fracturing or deep injection of waste fluids [129].

In order to determine the phase response curves, one does not have to carry out an explicit phase reduction of the underlying systems, but may rather focus on the occurrence of characteristic events. The latter are associated to large spikes of velocity of the block and are representative of earthquakes within the given models. With this in mind, one may effectively determine the phase response curves in complete analogy to the method typically applied for systems of spiking neurons. In particular, the impact of a perturbation is such that it locally changes the oscillation period of a system from the default value T_0 (period in the absence of perturbation) to a different value T_1. This may be used to numerically determine the phase shift $\Delta\varphi$ by measuring the relative change of the period [18, 55, 130]:

$$\Delta\varphi(\varphi) = \frac{T_0 - T_1}{T_0} \tag{23}$$

The phase shift $\Delta\varphi$ plotted as a function of the phase φ when a perturbation has kicked in constitutes precisely the phase response curve. If $T_1 < T_0$, the stimulus advances the cycle and vice versa. The change of period of the oscillation cycle where a perturbation has occurred defines the first-order phase response curve. Perturbations may also affect the duration of the next oscillation cycle T_2, which corresponds to the second-order phase response curve, where the phase shift is given by:

$$\Delta\varphi^{(2)}(\varphi) = \frac{T_0 - T_2}{T_0} \tag{24}$$

In the seismological context, a nontrivial profile of the second-order phase response curve would qualitatively reflect the long-term effects exerted on fault dynamics by the pulse perturbation, bearing in mind that the interseismic periods typically comprise very long time scales.

By implementing the described method, Franović et al. [48] determine the first- and second-order phase response curves for different models of fault dynamics. Apart from a single pulse perturbation, they also consider scenarios where two subsequent pulses are introduced within a given oscillation cycle. The form of the perturbation involves the standard α function

$$f(t) = C * [(-1/t_f) * exp(-(t - t_p)/t_f) + (1/t_r) * exp(-(t - t_p)/t_r)]\Theta(t - t_p) \tag{25}$$

whereby the Heaviside Θ function is used for shifting along the time-axis. Naturally, the rise and decline characteristic times t_r and t_f are selected so that the perturbation maintains a narrow profile compared to the oscillation period ($t_r = 0.15$, $t_f = 0.4$), whereas C is kept sufficiently small so that the perturbation does not affect the amplitude of the underlying oscillations (C = 5). Note that in all the considered instances, zero phase is assigned to the maximum amplitude of the u variable, which makes a natural choice in the seismological interpretation, because it corresponds to the occurrence of the characteristic event (earthquake).

5.1.2. Phase Response Curves for a Single Pulse Perturbation

The profiles of the first- and second-order phase response curves are provided in Figure 15(a), whereby the phase values are expressed in units of π. An important point regarding Figure 15(a) is that the first-order curve shows a phase advancement in a narrow phase interval lying at a small distance from the earthquake event, which is attributed the phase $\varphi = 0$. Nevertheless, the external stimulus introduced at all the other points of the oscillation cycle has a retardation effect, viz. it acts to postpone the next characteristic event. The change of sensitivity to a perturbation is expectedly found close to the end of the seismic cycle. In that phase domain, the delay effect is weaker, but the perturbation still cannot advance the cycle. Franović et al. [48] have verified that the characteristic profile of the phase response curve is not modified under variation of the perturbation amplitude within the relevant range of values [48].

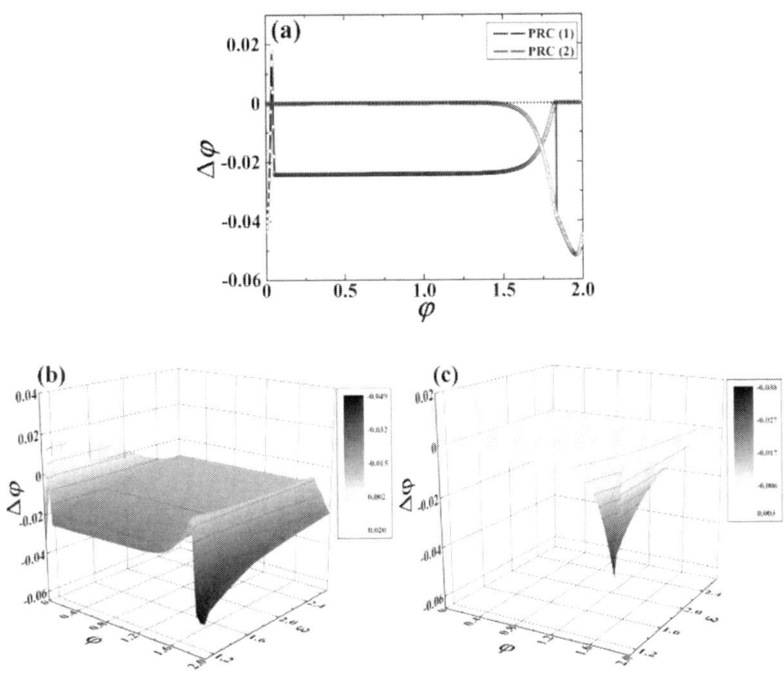

Figure 15. (a) Phase response curves of the first (blue circles) and second order (red circles) for a monoblock fault in the stick-slip regime. The fault parameters are (ε, ξ, γ) = (1.5, 0.5, 1000). (b) and (c) respectively show the families of first- and second-order PRCs $\Delta\varphi(\varphi, \varepsilon)$ for a monoblock fault under variation of ε. The remaining parameters are the same as in (a).

As to be expected, the second order curve in Figure 15 (a) corroborates that the perturbation typically has a negligible impact on the duration of the next seismic cycle. Nevertheless, an interesting point concerns a long-term retardation effect exhibited for $\varphi \approx 1.8\pi$. Note that this delaying effect occurs precisely in the phase domain where the first-order phase response curves show a reduced retardation.

An important issue is to examine how robust are the obtained phase response curve profiles against variation of fault parameters. Note that the parameters ξ and γ are less specific to particular faults, so that the impact of their variation may be of lesser significance compared to the one of changing ε, which is highly specific to particular faults [48]. An important finding is that the curve profiles from Figure 15 turn out to be generic, viz. they remain qualitatively unaffected by changing ξ or γ for fixed ε. The effects of varying ε under fixed ξ and γ are illustrated in Figure 15 (b) and Figure 15 (c). These figures refer to phase shifts corresponding to first- and second-order phase response curves respectively, whereby ε attains values from the seismologically relevant range $\varepsilon \in (1, 3)$, while ξ and γ are fixed at values from Figure 15 (a).

Regarding the first-order curve, the effect of advancing the seismic cycle by a perturbation introduced within a preferred time interval just after the earthquake persists for the most part of the relevant ε range, but declines with increasing ε. In fact, one also finds a critical ε value above which there is no phase advance, cf. Figure 15 (b). The other interesting effect, which consists in a reduced phase delay if the perturbation occurs close to the end of the seismic cycle, appears unaffected by variation of ε. Also, the delay effect characteristic for the most of phase domain becomes less pronounced with increasing ε. Therefore, the profile of the first-order curve in general becomes more flat as ε is enhanced. Similar may be stated in case of the second-order phase response curve. In particular, Figure 15 (c) clearly shows that the delay effect associated to perturbation occurring by the end of the oscillation cycle is gradually lost with ε.

5.1.3. Two-Pulse Phase Response Curves for a One-Block Fault

Franović et al. [48] further consider the sensitivity of a one-block fault in the regime of stick-slip oscillations to two successive pulse perturbations. The first pulse acts at the phase φ_1 of the oscillation cycle, whereas the other pulse is introduced at the phase φ_2. In both instances, the form of applied perturbation is assumed to be the same. Under real conditions, the occurrence of multiple perturbations during a single oscillation cycle may be attributed to a number of different phenomena, both natural and artificial.

The first-order phase response curve is shown in Figure 16 (a). Note that the term "phase response curve" is kept for simplicity, though the plot actually shows the dependence of a phase reset $\Delta\varphi$ in terms of φ1 and φ_2-φ_1. The same terminology is maintained when describing the analogous three-dimensional plots below. From Figure 16 (a) one learns of two different mechanisms that may give rise to advancement of the phase of the seismic cycle [48]. First, if the initial pulse is applied in a narrow interval sufficiently close to the last seismic event ($\varphi = 0$), the fault's phase is substantially advanced, irrespective of the precise point when the second perturbation occurs. Nonetheless, the advancing effect of two pulses is amplified compared to that of a single pulse, cf. Figure 16. The second mechanism is associated to the domain of φ_1 values away from the characteristic event. There, the earlier arrival of the first perturbation typically requires a late arrival of the second perturbation in order to induce a substantial phase advancement. However, for sufficiently large φ_1, the phase of seismic cycle is advanced only within a narrow interval of preferred φ_2 values, where the second pulse arrives in a relatively close succession to the first one. Outside of the described (φ_1, φ_2) domains, the impact of two successive pulse perturbations consists in delaying the next characteristic event, viz. the perturbations have a stabilizing effect on the fault dynamics.

Regarding the second-order phase response curve shown in Figure 16 (b), one notes a sizeable long-term effect if the first pulse arrives early (small φ_1), and the second pulse is applied sufficiently late within the given cycle (large φ_2-φ_1). It is interesting that the long-term effect may either result in fault destabilization (advanced phase of oscillation) or fault stabilization (delayed phase of oscillation), which depends sensitively on the phase of the second pulse. Note that we interpret phase advancement (retardation) of the seismic cycle as destabilization (stabilization) of the fault because its next characteristic event is precipitated (postponed) by the perturbation. The presence of both types of behavior is quite distinct from what is found for a single pulse perturbation in Figure 16 (b), where the only pronounced effect consists in delaying the next oscillation cycle.

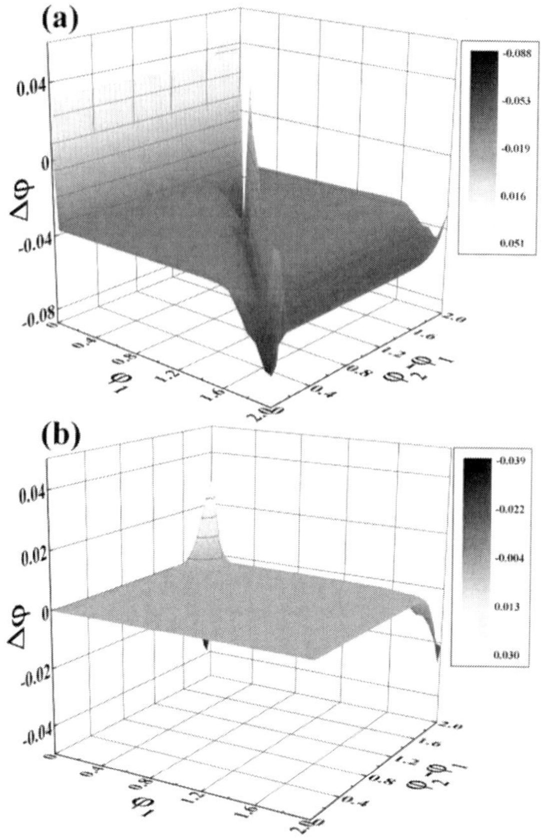

Figure 16. (a) and (b) respectively show the phase response curves of first and second order when a mono-block fault is subjected to two successive pulse perturbations. The first perturbation is introduced at the phase φ_1, and the second one is applied with the phase difference φ_2-φ_1. The fault parameters are set to $\varepsilon = 1.5$, $\xi = 0.5$, $\gamma = 1000$.

5.2. Phase Response of Complex Faults

This subsection concerns the behavior of complex faults, whose structure generally involves multiple segments with potentially different elastic and frictional properties.

In particular, Franović et al. [48] analyze the sensitivity to perturbation of a homogeneous fault ($\varepsilon_1 = \varepsilon_2$, $\xi_1 = \xi_2$, $\gamma_1 = \gamma_2$) built of identical blocks, as well as the heterogeneous complex fault [48]. Heterogeneity is confined to the scenario of two blocks with disparate ε, $\varepsilon_1 \neq \varepsilon_2$, which results in distinct periods of the respective stick-slip oscillations. Both for the homogeneous and the heterogeneous two-segment faults, *c* is taken sufficiently weak so that the interaction does not perturb the respective oscillation cycles of the blocks.

5.2.1. Phase Response Curves for the Fault Comprised of Two Identical Blocks

For the homogeneous complex fault, Franović et al. [48] analyze the scenario where perturbation on block 1 acts at phase φ_1 of its oscillation cycle, whereas block 2 receives a kick with the phase difference $\varphi_2 - \varphi_1 > 0$. The form of perturbation to both blocks is assumed to be identical.

The first- and second-order phase response curves are illustrated in Figure 17. The respective phase shifts are denoted by $\Delta\varphi_{ij}$, where the first index denotes a particular block, and the second index refers to the first/second order of the phase response. It is interesting to compare the first-order phase response curves in Figure 17 (a) and Figure 17 (c), because this indicates how the interaction affects the response of individual blocks. In particular, the phase of both blocks is significantly advanced if the first block is stimulated immediately after the characteristic event. In this case, a perturbation of the first block induces a strong destabilization of the second block, regardless of the moment when the second block is perturbed. Just beyond the described region of φ_1 values, one encounters a narrow domain where the external stimuli delay the cycles of both blocks. Nevertheless, the most interesting point concerns the differences between $\Delta\varphi_{11}$ and $\Delta\varphi_{21}$ dependencies. Namely, Franović et al. [48] have found that $\Delta\varphi_{21}$ shows a much larger (φ_1, φ_2) domain where the phase of the cycle is strongly advanced compared to $\Delta\varphi_{11}$. This point corroborates that the dynamics of block 2 is substantially affected by the perturbation of block 1 conveyed via the interaction term. In fact, within the indicated (φ_1, φ_2) domain, the destabilization effect on block 2 is amplified by the co-action of two pulses, reflected in an indirect influence of a perturbation applied to the first block, and a direct impact of the subsequent pulse. Note that the destabilization effect on block 2 is more pronounced if the perturbation on block 1 arrives by the end of its oscillation cycle.

As far as the second order phase response curves are concerned, Figure 17 (b) and Figure 17 (d) both show quite large (φ_1, φ_2) domains of substantial phase advancement and phase retardation. Such nontrivial long-term effects are caused by the interaction between the blocks. Nevertheless, note that for the same (φ_1, φ_2) values, the long-term effects on two blocks are asymmetric, in a sense that the stabilization (phase delay) of one block is accompanied by the destabilization (phase advancement) of the other block.

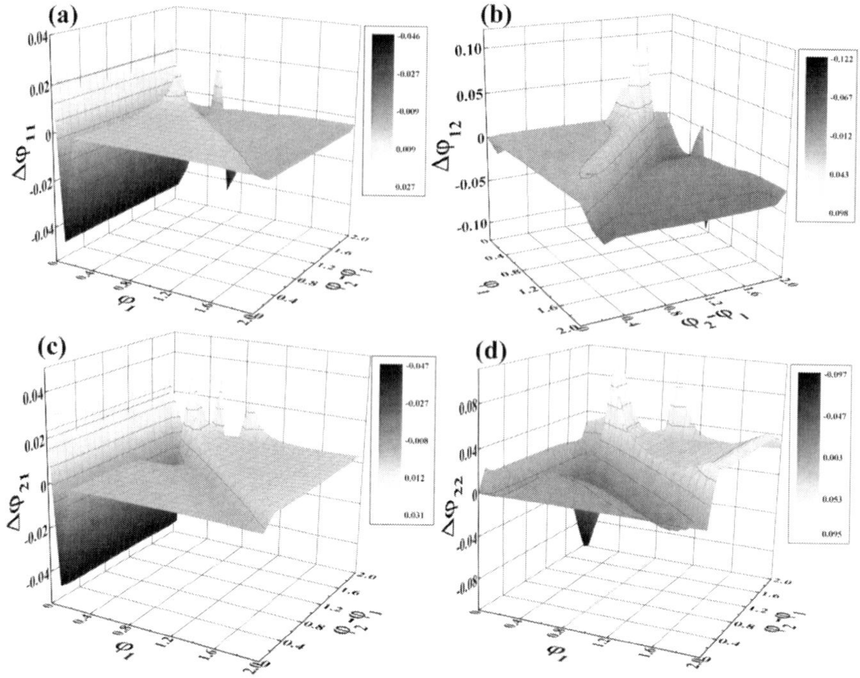

Figure 17. The top (bottom) row shows the phase responses of the first (left column) and second order (right column) for block 1 (2) in dependence of φ_1 and φ_2-φ_1. The blocks are assumed to be identical, and are characterized by parameters $\varepsilon = 1.5$, $\xi = 0.5$, $\gamma = 1000$. The interaction strength $c = 0.1$ lies well below the critical bifurcation value and warrants that the periodic oscillations on the coupled blocks are not substantially different from those in the uncoupled case (For interpretation of the references to colour in this figure legend, the reader is referred to the web version of this chapter).

5.2.2. Phase Response Curves for the Two-Block Inhomogeneous Fault

Franović et al. [48] further determine the phase response curves for an inhomogeneous fault comprised of two blocks with distinct ε values. The latter are such that the respective oscillation periods of blocks are substantially different, with $T_1 \approx 52$ and $T_2 \approx 77$. Two generic scenarios are considered: in the first instance, perturbation is applied only to the block with the shorter oscillation period (here block 1), whereas in the second instance one perturbs the block with the longer oscillation period (here block 2). The numerical experiments are carried out so that at the moment when the stimulus arrives to one block, the other block always attains the same phase.

The results for the first scenario (perturbation applied to block 1 at phase φ_1) are illustrated in Figure 18, whereby Figure 18 (a) and Figure 18 (b) refer to first- and second-order responses of blocks 1 and 2, respectively. Note that Figure 18 (c) shows the average responses $\Delta\Phi_i = (\Delta\varphi_{1,i} + \Delta\varphi_{2,i})/2$ for the total system (complex fault), where $i \in \{1, 2\}$ stands for the first- or second-order response.

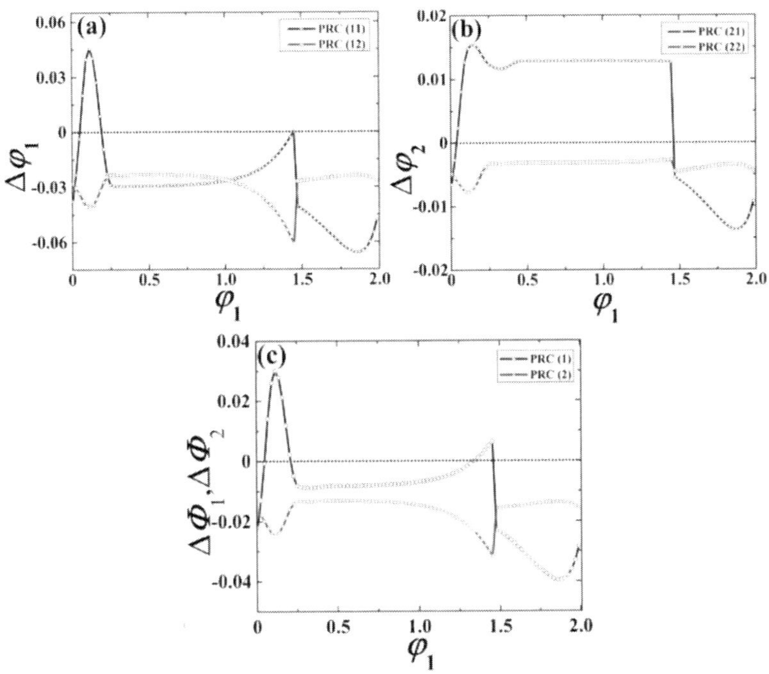

Figure 18. Scenario where pulse perturbation is introduced to the block with shorter oscillation period. In (a) are shown the first- (blue circles) and second-order phase response curves (red squares) for block 1, which is subjected to pulse perturbation. (b) illustrates the first- and second-order phase response curves for block 2 which is influenced by perturbation only via interaction with block 1. (c) provides an indication on the average phase response $\Delta\Phi_i$, $i \in (1,2)$ for the total system, viz. the complex fault, whereby index *i* refers to the first- or second-order dependence. The block parameters kept fixed are $\xi = 0.5$, $\gamma = 1000$, whereas ε values on particular blocks are $\varepsilon_1 = 1.45$ and $\varepsilon_2 = 2$ (For interpretation of the references to colour in this figure legend, the reader is referred to the web version of this chapter).

As expected, for block 1, the first- and second-order phase response curves are qualitatively similar to that of an uncoupled block, cf. Figure 18 (a). As far as block 2 is concerned, note that Figure 18 (b) shows the dependence

$\Delta\varphi_2(\varphi_1)$ obtained for a fixed value of the phase of the second block. In other words, a perturbation is applied at different phases of the cycle of block 1, whereas block 2 at the moment of pulse arrival to block 1 always acquires certain fixed phase φ_2. The first-order response of block 2 implies that the interaction may play an important role in destabilization of the fault. In particular, it strikes that a perturbation acting on the block with a shorter oscillation period (block 1) is found to substantially advance the phase of the block with the longer oscillation period (block 2) for a broad interval of φ_1 values [48]. Note that Figure 18 (c) implies that the average response of the two-block fault is mostly shaped by the behavior of block 1 which actually receives the stimulus.

In analogy to the previous setup, perturbation is applied at different phases of the cycle of block 2, whereas block 1 at the moment of pulse arrival to block 2 always has a fixed phase φ_1. The first-order responses of the blocks are shown in Figure 19 (a), whereas Figure 19 (b) refers to the second-order responses. The average first- and second-order responses of the complex two-block fault are provided in Figure 19 (c).

At variance with the scenario from Figure 18, the first-order phase response curve for the kicked block now shows two phase intervals which admit an advancement of the seismic cycle, whereby one is closely after the characteristic event ($\varphi_2 \approx 0.1\pi$), while the other is located by the end of the seismic cycle ($\varphi_2 \approx 1.5$). Nevertheless, an important qualitative finding on the first-order response of block 1 is that for almost all φ_2, the perturbation on block 2 advances the oscillation cycle on block 1. The analogous effect of advancing the phase of the respective block which is not subjected to pulse perturbation has already been observed in Figure 18 (b), but not in such a broad domain of perturbation phases. This point clearly evinces that (i) the effect of perturbation conveyed via interaction between the blocks is non-negligible, and (ii) that its impact on the block that has not directly received the pulse perturbation is typically destabilizing. As far as the total system is concerned, the first-order response is mostly influenced by the behavior of the block explicitly affected by the perturbation, whereas the leading delay effect in the second-order response derives from the other block, cf. Figure 18 (c).

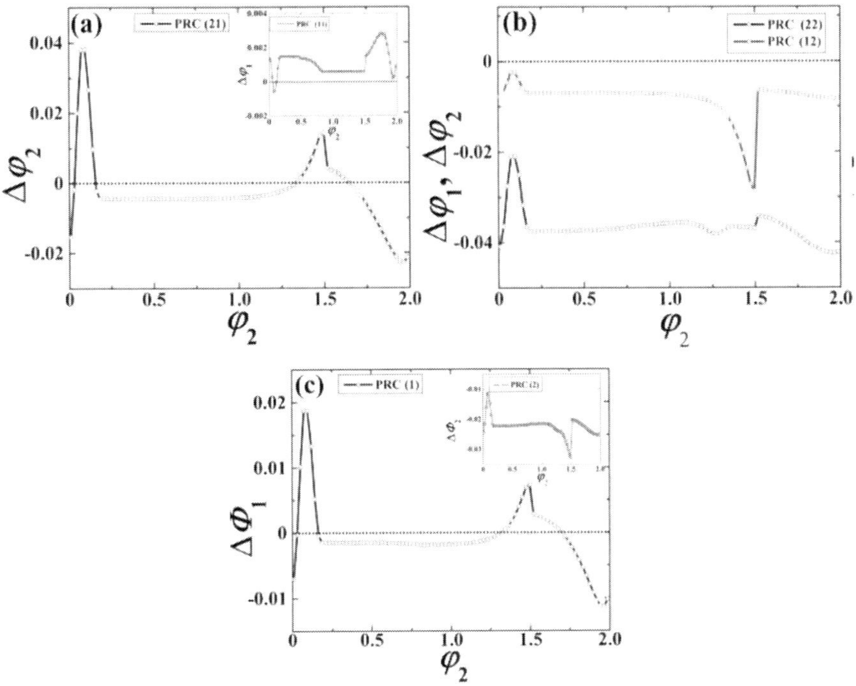

Figure 19. Scenario where pulse perturbation acts on the block with longer oscillation period. The main frame and inset in (a) shows the first-order phase response curve for block 2 and block 1, respectively. In (b) are shown the second-order phase response curves, whereby the blue circles (red squares) are reserved for block 2 (block 1). The main frame and the inset in (c) illustrate the average first- and second-order phase response for the complex fault, respectively. The block parameters are the same as in Figure 18 (For interpretation of the references to colour in this figure legend, the reader is referred to the web version of this chapter).

CONCLUSION

In this chapter, authors provide a brief review of the main results regarding the dynamics of one-dimensional spring-block models under the impact of the following controlling factors:

- Delayed interaction among the blocks within array
- Delayed friction response to the change in velocity
- Incoherent random noise
- Various friction strength

- Continuous and transient perturbations.

This chapter represents the retrospective of the authors' work, since it is mainly based on the previous studies by Kostić et al [19, 20, 22, 61, 62], Franović et al. [48] and Vasović et al. [21]. Different setups of spring-block models, including a mono-block model, two-block model and an array of blocks, are derived from the original Burridge-Knopoff model, which is comprised of a series of interconnected blocks that are moving in a stick-slip fashion along the rough plane, in which way they mimic the seismogenic fault motion. Dynamics of such models is analyzed using standard method of nonlinear dynamics: bifurcation analysis, mean-field method, construction of phase response curves and numerical analysis. Such an approach that is based on the application of techniques from theoretical physics to geophysical research is new for the seismological community, and it provides a new perspective on the qualitative analysis of the earthquake nucleation process.

Previous studies aim to explain the very nature of the complex earthquake fault motion, starting from the simple dynamical systems, which assume that the main nonlinearity comes from the specific friction between the moving blocks and the rough plane, including rate-dependent and rate – and state – dependent friction law. However, if no additional assumption is made, analyzes show that such models are incapable of reproducing any relevant dynamical features. On the other hand, if supplementary presumption are made, including the delayed response of friction on the velocity change, delayed interaction among the block, the effect of random seismic noise or the influence of continuous or transient perturbation, one could easily detect rich dynamical behavior of these simple starting models. New features that arise include the following:

- direct or inverse Hopf bifurcations, with transition from equilibrium state (fixed point), through periodic (limit cycle) and quasiperiodic (torus) oscillations to long-term aperiodic irregular behavior (deterministic chaos), giving rise to the occurrence of complex fault motion;
- amplitude or oscillation death, preventing the occurrence of complex fault motion;
- advancement or retardation of the seismic cycle, in dependence on the nature of the assumed perturbation.

Assumption of delayed friction response to the velocity change results in two possible scenarios. Depending on the values of the control parameters, time delay could act both as a stabilizing and destabilizing factor. In the latter case, increase of time delay in friction could induce the more complex dynamics of fault motion, eventually leading to the onset of deterministic chaos. In the former case, increase of time delay could also lead to amplitude or oscillation death, acting, in that way, as a stabilizing factor.

If one assumes transient perturbation of a velocity in a form of sine wave scaled by an exponential pulse, which could correspond to the effect of a passing seismic wave from near or distant earthquake, transitions to different seismic regimes could be observed as a function of the assumed duration, frequency and the perturbation moment. In particular, if high-frequent acceleration changes are presumed in the inter-seismic creep-regime, this could lead to quasiperiodic-like dynamics, while long-period changes of acceleration in both post-seismic and inter-seismic creep regimes induce chaos-like behavior, which can be treated as an onset of the seismic motion.

If change of control parameters is assumed in a form of continuous periodical regular wave, which could correspond to the regular periodic influence of mining machines, then a direct transition from periodic motion to deterministic chaos could be observed in dependence on the assumed perturbation frequency. Moreover, the obtained results indicate the predominant impact of the assumed perturbation amplitude, which acts a controlling factor for the occurrence of a long-term irregular aperiodic behavior.

Except from introducing new parameters to the existing models, which triggered the occurrence of new dynamical regimes, previous studies on this topic also included the application of mean-field approach, which turned out to be a very useful technique when analyzing stochastic dynamical systems with large number of units. In particular, this method enables derivation of a simple deterministic model governed by a small number of equations, which exhibits qualitatively the same dynamical features as the starting stochastic model with a large number of equations. In this way, one could discuss about the conditions for specific transitions of the complex system by analyzing the reduced deterministic system.

Also, authors present the analysis of the perturbation by applying the theory of phase response curves, which provided more extended picture of the changes in dynamics of a mono-block and two-block models under the influence of external perturbation. The resulting first-order and second-order phase response curves imply the occurrence of seismic cycle advancement or delay in dependence on the model setup (mono-block, homogeneous or heterogeneous two-block model) and on the arriving time of the presumed single-pulse or two-pulse stimulus.

In relation to seismology, it is interesting to note that the observed dynamics of the analyzed spring-block models, in certain cases, exhibit power-law behavior. In particular, if the sum of displacements of all the blocks in an array is treated as analogous to the earthquake magnitude, one could easily show that distribution of different events, regarding their number and size, is accurately described by Gutenberg-Richter law, with near $b \approx 1$ value.

The presented analysis of the dynamics of the spring-block models enables the following:

- determination of the key parameters that control the system dynamics,
- qualitative assessment of the effect of various parameters on the changes of dynamical regimes,
- easier inquiry of the complex stochastic model by deriving the reduced, simpler deterministic model with qualitatively the same dynamical features as the stochastic model.

The main limitation of the presented approach lies in definition of the starting model, which is a dimensionless representation of the real fault motion. This means in the analysis one deals with parameter values which strictly do not correspond to the real observed data. Concerning this, one is able to infer about the qualitative influence of the parameter on the dynamics of the observed system, i.e., whether the increase or decrease of parameter values affect the system's dynamics, and how subtle those changes need to be.

Further research of the dynamics of spring-block models could include the analysis of two-dimensional models, which, besides temporal, also include the spatial variations and irregularities. Such models are definitely closer to the real observed fault motion, and, therefore, could more closely capture the specifics of movement along the fault plane. On the other hand, future analysis could also be oriented towards the application of different techniques of nonlinear time series analysis in order to capture the main dynamical mechanism behind the strong ground motion. Such study could result in useful data for engineering design, by providing the more accurate mathematical description of earthquake ground motion, or by determining the critical parameter values for which strong irregular ground motion occurs. Another approach that could be also implemented concerns the analysis of the effect of normal and shear stress on the earthquake nucleation process, including both the normal (reverse) and strike-slip faults.

ACKNOWLEDGMENTS

The work on this chapter was partly supported by the Ministry of Education, Science and Technological Development of the Republic of Serbia (Contracts No. 176016 and 171017).

REFERENCES

[1] Hess, D. P., Soom, A. *J Tribol*, 1990, 112, 147 - 152.
[2] Burić, N., Mudrinić, M., Vasović, N. *Chaos Soliton Fract*, 2001, 12, 483-489.
[3] Burić, N., Grozdanović, I., Todorović, K., Vasović, N. *J Stat Phys,* 2011, 145, 175-186.
[4] Burić, N., Ranković, d., Todorović, K., Vasović, N. *Physica A,* 2010, 389, 3956-3964.
[5] Burić, N., Grozdanović, I., Vasović, N. *Chaos Soliton Fract*, 2004, 22, 731-740.
[6] Burić, N., Grozdanović, I., Vasović, N. *Chaos Soliton Fract*, 2005, 23, 1221-1233.
[7] Burić, N., Vasović, N. *Chaos Soliton Fract*, 2007, 31, 336-342.
[8] Vasović, N., Burić, N., Todorović, K., Grozdanović, I. *Chinese Phys B,* 2012, 21, 010203.
[9] Burić, N., Todorović, K., Vasović, N. *Physical Review E*, 2010, 82, 037201.
[10] Burić, N., Todorović, K., Vasović, N. *Chaos Soliton Fract*, 2009, 40, 2405-2413.
[11] Burić, N., Todorović, K., Vasović, N. *Phys Rev E,* 2007, 75, 026209.
[12] Burić, N., Todorović, K., Vasović, N. *Chaos Soliton Fract*, 2009, 40, 1127-1135.
[13] Burić, N., Todorović, K., Vasović, N. *Phys Rev E,* 2008, 78, 036211.
[14] Burić, N., Todorović, K., Vasović, N. *Phys Rev E,* 2007, 75, 067204.
[15] Kuramoto, Y. Chemical Oscillations, Waves, and Turbulence; Dover, Mineola: New York, USA, 2003; 176 p.

[16] Winfree, A. T. The Geometry of Biological Time; Springer: Berlin, 1980; 779 p.
[17] Tass, P. A. Phase Resetting in Medicine and Biology: Stochastic Modeling and Data Analysis; Springer: Berlin Heidelberg, 2007; 329 p.
[18] Schultheiss, N. W., Prinz, A. A., Butera, R. J. (Eds.). Phase Response Curves in Neuroscience: Theory, Experiment, and Analysis; Springer: New York, 2012; 518 p.
[19] Kostić, S., Franović, I., Todorović, K., Vasović, N. *Nonlinear Dynam,* 2013, 73, 1933-1943.
[20] Kostić, S., Franović, I., Perc, M., Vasović, N., Todorović, K. *Sci Rep: Nature Publishing Group*, 2014, 4, 5401.
[21] Vasović N., Kostić S., Franović I., Todorović K. *Commun Nonlinear Sci,* 2016, 38, 117-129.
[22] Kostić, S., Vasović, N., Franović, I., Todorović, K., Klinshov, V., Nekorkin, V. Submitted to Nonlinear Dynam, 2016 (under review).
[23] Reid, H. F. Report of the State Investigation Commission, Vol. 2, Carnegie Institution of Washington, Washington, D. C. 1910, pp. 16-28.
[24] Scholz, C. H. The mechanics of earthquakes and faulting, Cambridge University Press: Cambridge, 2002; 504 p.
[25] Bykov, V. G. *Acta Geophys*, 2008, 56, 270-285.
[26] Brace, W., Byerlee, J. *Science,* 1966, 153, 990–992.
[27] Burridge, R., Knopoff, L. *Bull. Seismol. Soc. Am.* 1967, 57, 341–371.
[28] Kostić, S., Vasović, N. *Tehnika*, 2011, 67, 61-66.
[29] Omori, F. J Coll Sci, Imperial University of Tokyo, 1894, 7, 111–200.
[30] Utsu, T. *Geophysical Magazine*, 1961, 30, 521–605.
[31] Kawamura, H., Hatano, T., Kato, N., Biswas, S., Chakrabarti, B. K. *Rev Mod Phys*, 2012 84, 839-884.
[32] Carlson, J. M., Langer, J. S. *Phys Rev A*, 1989, 40, 6470–6484.
[33] De Sousa Vieira, M. *Phys Rev Lett*, 1999, 82, 201-204.
[34] Erickson, B., Birnir, B., Lavallee, D. *Nonlinear Proc Geoph*, 2008, 15, 1-12.
[35] Erickson, B. A., Birnir, B., Lavallée, D. *Geophys J Int*, 2011, 187, 178-198.
[36] Dieterich, J. H. *J Geophys Res*, 1979, 84, 2161–2168.
[37] Ruina, A. L. *J Geophys Res*, 1983, 88, 10359-10370.
[38] Clancy, I., Corcoran, D. *Phys Rev E,* 2009, 80, 016113.
[39] Scholz, C. H. *Nature,* 1998, 391, 37-42
[40] Tullis, T. E., Beeler, N. M., Weeks, J. D. Eos, *Trans Am Geophys Union,* 1993, 74, 296.

[41] Beeler, N. M., Tullis, T. E., Weeks, J. D. *Geophys Res Lett*, 1994, 21, 1987–1990.
[42] Perrin, G., Rice, J. R., Zheng, G. *J Mech Phys Solids*, 1995, 43, 1461–1495.
[43] Scholz, C. H., Molnar, P., Johnson, T. *J Geophys Res*, 1972, 77, 6392–6406.
[44] Baumberger T, Heslot F, Perrin B. *Nature,* 1994, 367, 544–546.
[45] Marone, C. Annu Re. *Earth Planet Sci*, 1998, 26, 643–696.
[46] Linker, M. F., Dieterich, J. H. *J Geophys Res*, 1992, 97, 4923–4940.
[47] Wang, W., Scholz, C. H. *Pure Appl Geophys*, 1994, 143, 303–316.
[48] Franović, I., Kostić S., Perc, M., Klinshov, V., Nekorkin, V., Kurths, *J. Chaos,* 2016, 26, 063105-1-12.
[49] Scholz, C. H. *Bull Seismol Soc Am*, 2010, 100, 901–909.
[50] Sugiura, N., Hori, T., Kawamura, Y. *Nonlin Proc Geophys*, 2014, 21, 251-267.
[51] Lursmanashvili, O., Paatashvili, T., Gheonjian, L., in Synchronization and Triggering: from Fracture to Earthquake Processes; Editors, de Rubeis, V., Czechowski, Z., Teisseyre, R., Springer-Verlag: Berlin-Heidelberg, 2010, pp. 305-322.
[52] Vasudevan, K., Cavers, M., Ware, A. *Nonlin Proc Geophys*, 2015, 22, 499–512.
[53] Perez Velazquez, J. L., Galan, R. F., Garcia Dominguez, L., Leshchenko, Y., Lo, S., Belkas, J., Guevara Erra, R., *Phys Rev E,* 2007, 061912.
[54] Tateno, T., Robinson, H. P. C. *Biophys J,* 2007, 92, 683–695.
[55] Krishnan, G. P., Bazhenov, M., Pikovsky, A. *Phys Rev E*, 2013, 88, 042902.
[56] Achuthan, S., Canavier, C. C. *J Neurosci*, 2009, 29, 5218-5233.
[57] Kawamura, Y., Nakao, H., Arai, K., Kori, H., Kuramoto, Y. *Phys Rev Lett,* 2008, 101, 024101.
[58] Kori, H., Kawamura, Y., Nakao, H., Arai, K., Kuramoto, Y. *Phys Rev E,* 2009, 80, 036207.
[59] Ko, T. W., Ermentrout, G. B. *Phys Rev E,* 2009, 79, 016211.
[60] Levnajić, Z., Pikovsky, A. *Phys Rev E,* 2010, 82, 056202.
[61] Kostić, S., Vasović, N., Franović, I., Todorović, K. *Nonlin Proc Geophys,* 2013, 20, 857-865.
[62] Kostić, S., Vasović, N., Franović, I., Todorović, K. *J Comput Nonlin Dyn,* 2013, 9, 031019.

[63] Prasad, A., Dhamala, M., Adhikari, B. M., Ramaswamy, R., *Phys Rev E,* 2010, 81, 027201.
[64] Newhouse, S., Ruelle, D., Takens, F. *Commun Math Phys*, 1978, 64, 35–44.
[65] Bizzari, A. *Earth-Sci Rev, 2012*, 115, 304–318.
[66] Day, S. M., Dalguer, L. A., Lapusta, N., Liu, Y. *J Geophys Res*, 2005, 110, B12307.
[67] Bizzarri, A., Belardinelli, M. E. *Geophys J Int*, 2008, 173, 906–921.
[68] Rubin, A. M., Ampuero, J. P. *J Geophys Res*, 2005, 110, B11312.
[69] Bizzarri, A., Spudich, P. *J Geophys Res*, 2008, 113, B05304.
[70] Brodsky, E. E., Karakostas, V., Kanamori, H. *Geophys Res Lett*, 2000, 27, 2741–2744.
[71] Brodsky, E. E., Prejean, S. G. *J Geophys Res*, 2005, 110, B04302.
[72] Freed, A. M. *Annu Rev Earth Planet Sci*, 2005, 33, 335–367.
[73] Hill, D. P. et al. *Science,* 1993, 260, 1617–1623.
[74] Prejean et al. *Bull Seismol Soc Am*, 2004, 94, 348–359.
[75] Van Der Elst, N. J., Brodsky, E. E. *J Geophys Res*, 2010, 115, B07311.
[76] Van Der Elst, N. J. The effect of seismic waves on earthquake nucleation and fault strength. University of California, Santa Cruz, PhD dissertation, 2012; 210 p.
[77] Gomberg, J., Blanpied, M. L., Beeler, N. M. *Bull Seismol Soc Am,* 1997, 87, 294–309.
[78] Gomberg, J., Beeler, N. M., Blanpied, M. L., Bodin, P. *J Geophys Res,* 1998, 103, B10, 24411–24426.
[79] Belardinelli, M. E., Bizzarri, A., Cocco, M. *J Geophys Res,* 2003, 108, 2135.
[80] Gomberg, J., Bodin, P., Savage, W., Jackson, M. E., *Geology,* 1995, 23, 41–44.
[81] Perfettini, H., Schmittbuhl, J., Rice, J. R., Cocco, M. *J Geophys Res*, 2001, 106, 13455–13472.
[82] Sone, H., Shimamoto, T. *Nat Geosci*, 2009, 2, 705-708.
[83] Lapusta N.: The roller coaster of fault friction. *Nat Geosci*, 2009, 2, 676-677.
[84] Lienkaemper, J. J. et al. *Bull Seismol Soc Am*, 2012, 102, 31-41.
[85] Galehouse, J. S., Lienkaemper, J. *Bull Seismol Soc Am*, 2004, 93, 2415-2433.
[86] Parsons, T., Console, R., Falcone, G., Muru, M., Yamashina, K. *Geophys J Int*, 2012, 190, 1673-1688.
[87] Franović, I., Miljković, V. *Eur Phys J B*, 2010, 763, 613-624.

[88] Franović, I., Miljković, V. *Chaos Soliton Frac*, 2011, 44, 122-130.
[89] Franović, I., Miljković, V. *Commun Nonlin Sci,* 2011, 16, 623-633.
[90] Franović, I., Miljković, V. EPL 92, 2010, 68007.
[91] Telford, W. M., Geldart, L. P., Sheriff, R. E. Applied Geophysics; Cambridge University Press: Cambridge, 1990; 792 p.
[92] Ryabov, V. B., Correig, A. M., Urquizu, M., Zaikin, A. A. *Chaos Soliton Fract,* 2003, 16, 195-210.
[93] Kanamori, H. (Ed.) Earthquake Seismology: Treatise on Geophysics; Elsevier: Amsterdam, 2009; Volume 4, 653 p.
[94] Bhattacharyya, P., Chakrabarti, B. (Eds.), Modeling Critical and Catastrophic Phenomena in Geoscience; Springer: Berlin Heidelberg, 2006; 527 p.
[95] Wesnousky, S. G. *Bull Seismol Soc Am*, 1994, 84 1940-1959.
[96] Sieh, K. *Proc Natl Acad Sci USA*, 1996, 93, 3764 – 3771.
[97] Tapponnier, P., Ryerson, F. J., Van der Woerd, J., Meriaux, A. S., Lasserre, C. C R *Acad Sci,* 2001, 333, 483 – 494.
[98] Nishenko, S. P., Buland, R. *Bull Seismol Soc Am*, 1987, 77, 1382-1399.
[99] Parsons, T. *Geophys Res Lett,* 2008, 35 L21301.
[100] Sykes, L. R., Menke, W. B *Seismol Soc Am*, 2006, 96, 1569-1596.
[101] Ishibashi, K. *Ann Geophys*, 2004, 47, 339-368.
[102] Benedetti, L. et al. *J Geophys Res Solid Earth*, 2013, 118, 4948-4974.
[103] Scharer, K. M., Biasi, G. P., Weldon, R. J., Fumal, T. E. *Geology*, 2010, 38, 555-558.
[104] Klinger, Y., Etchebes, M., Tapponnier, P., Narteau, C. *Nat Geosci*, 2011, 4, 389-392.
[105] Haibing, L., Van der Woerd, J., Tapponnier, P., Klinger, Y., Xuexiang, Q., Jingsuia, Y., Yintang, Z. *Earth Planet Sci Lett*, 2005, 237, 285–299.
[106] Matsuzawa, T., Igarashi, T., Hasegawa, A. *Geophys Res Lett,* 2002, 29, 1543.
[107] Carlson, J. M. *Phys Rev A,* 1991, 44, 6226.
[108] Mori, T., Kawamura, H. *Phys Rev Lett*, 2005, 94, 058501.
[109] Mori, T., Kawamura, H. *J Geophys Res,* 2066, 111, B07302.
[110] Kotani, T., Yoshino, H., Kawamura, H. *Phys Rev E,* 2008, 010102(R).
[111] Sammis, C. G., Smith, S. W. *Tectonophysics,* 2013, 589, 167-171.
[112] Guckenheimer, J. *J Math Biol*, 1975, 1, 259-273.
[113] Schwabedal, J. T. C., Pikovsky, A. *Phys Rev Lett*, 2013, 110, 204102.
[114] Mirollo, R. E., Strogatz, S. H. *SIAM J Appl Math*, 1990, 50, 1645-1662.
[115] Ermentrout, G. B., Kopell, N. *J Math Biol*, 1991, 29, 195-217.
[116] Kuramoto, Y. *Physica D,* 1991, 50, 15-30.

[117] Bottani, S. *Phys Rev Lett*, 1995, 74, 4189.
[118] Glass, L. *Nature*, 2001, 410, 277-284.
[119] Glass, L., Nagai, Y., Hall, K., Talajić, M., Nattel, S. *Phys Rev E*, 2002, 65, 021908.
[120] Canavier, C. C., Achuthan, S. *Math Biosci*, 2010, 226, 77-96.
[121] Ermentrout, G. B. *Neural Comput*, 1996, 8, 979–1001.
[122] Ermentrout, G. B., Beverlin Bryce, I. I., Troyer, T., Netoff, T. *J Comput Neurosci*, 2011, 31, 185–197.
[123] Galan, R. F., Ermentrout, G. B., Urban, N. N. *Phys Rev Lett*, 2005, 94, 158101.
[124] Chouet, B. A. *Nature*, 1996, 380, 309-316.
[125] Belardinelli, M. E., Bizzarri, A., Cocco, M. *J Geo Res*, 2003, 108, 2135.
[126] Perfettini, H., Schmittbuhl, J., Cochard, A. *J Geophys Res*, 2003, 108(b9), 2409.
[127] Lu, W., Yang, J., Yan, P., Chen, M., Zhou, C., Luo, Y., Jin, L. *Int J Rock Mech Min Sci*, 2012, 53, 129-141.
[128] Mulargia, F., Bizzarri, A. *Sci Rep*, 2014, 4, 6100.
[129] Ellsworth, W. L. *Science*, 2013, 341, 1225942.
[130] Ermentrout, G. B., Terman, D. H. Mathematical Foundations of Neuroscience. Springer: New York, 2010; 422 p.

In: Earthquakes
Editor: Wayne Coleman

ISBN: 978-1-53610-342-7

Chapter 2

MODELING OF SEISMIC LIQUEFACTION DATA USING EXTREME LEARNING MACHINE

Pijush Samui[1,*], Sanjiban Sekhar Roy[†], Pradeep Kurup[3,‡] and Yıldırım Dalkiliç[4,§]

[1]Department of Civil Engineering,
NIT Patna, Bihar, India
[2]SCOPE, VIT University, Vellore, India
[3]Department of Civil and Environmental Engineering,
University of Massachusetts Lowell, Lowell, MA, US
[4]Faculty of Engineering,
Civil Engineering Department, Erzincan University, Turkey

ABSTRACT

Seismic liquefaction is a major concern for any earthquake prone area. This article employs a new technique based on Extreme Learning Machine (ELM) for determination of liquefaction susceptibility of soil based on Standard Penetration Test (SPT) and Cone Penetration Test (CPT) from the Chi-Chi earthquake. ELM is the modified version of

[*] Email: pijush.phd@gmail.com.
[†] Email: sanjibanroy09@gmail.com.
[‡] Email: Pradeep_Kurup@uml.edu.
[§] Email: yildirim.dalkilic@gmail.com.

Single-hidden Layer Feed forward Neural Network. Different input combinations have been tried to get best performance. A comparative study has been carried out between the SPT and CPT based models. CPT based models give better performance than the SPT based model. The developed ELM model shows that Cone Resistance (q_c) and Peck Ground Acceleration (PGA) are sufficient parameters for determination of liquefaction susceptibility of soil.

Keywords: extreme learning machine, liquefaction, standard penetration test, cone penetration test, prediction

INTRODUCTION

Liquefaction of soil is one of the main concerns for the stability of civil engineering structures. Therefore, the determination of seismic liquefaction potential of soil is an important task for designing civil engineering structures. Geotechnical engineers use different methods for determination of seismic liquefaction potential of soil (Seed and Idriss, 1967; Seed and Idriss, 1971; Seed et al., 1983; Seed et al., 1984; Youd et al., 2001; Seed and Idriss, 1982; Ambraseys, 1988; Arango, 1996; Andrus and Stokoe, 1997; Youd and Noble, 1997a; Robertson and Campanella, 1985; Skempton, 1986). Every method has own its advantages and disadvantages.

This article adopts Extreme Learning Machine (ELM) for determination of seismic liquefaction potential of soil based on Standard Penetration Test (SPT) and Cone Penetration Test (CPT) data from the Chi-Chi earthquake. ELM is modified version of Single hidden Layer Feed forward Networks (SLFN) (Huang et al., 2006). It does not suffer the effect of the curse of dimensionality (Wang et al., 2011). It determines output weights analytically. It has been successfully applied to many real world problems (Huang et al., 2006; Yeu et al., 2006; Horata et al., 2013; Zong et al., 2013). Four models have been developed. Table 1 shows the details of inputs of four models. The dataset has been collected from the Chi-Chi earthquake (Hwang and Yang, 2001; Ku et al., 2004). Four models have been developed. Table 1 shows the inputs of the four models. A comparative study has been carried out between the developed four models.

Table 1. Details of the models

Model	Inputs
I	CSR, q_c
II	PGA, q_c
III	CSR, N
IV	PGA, N

DETAILS OF ELM

The basic concept of ELM is taken from SLFN. Let us consider the following dataset (D).

$$D = \{(x_i, y_i)\}_{i=1}^{N} \tag{1}$$

where x is input, y is output and N is the number of dataset.

According to SLFN, the relation between x and y is given below.

$$\sum_{i=1}^{K} \beta_i g_i (w_i.x_j + b_i) = y_j \tag{2}$$

where w is weight between hidden layer and input layer, β is weight between hidden layer and output layer and b is threshold.

The above equation can be written in the following way.

$$H\beta = T \tag{3}$$

where $H = \begin{bmatrix} g(w_1.x_1 + b_1) & \cdots & g(w_K.x_1 + b_K) \\ \vdots & \ddots & \vdots \\ g(w_1.x_N + b_1) & \cdots & g(w_{K.}x_N + b_K) \end{bmatrix}$, $\beta = \begin{bmatrix} \beta_i^T \\ \vdots \\ \beta_K^T \end{bmatrix}_{K \times N}$ and $T = \begin{bmatrix} t_1^T \\ \vdots \\ t_N^T \end{bmatrix}$.

The value of β is determined from the following equation.

$$\beta = H^{-1}T \tag{4}$$

where H^{-1} is the Moore–Penrose generalized inverse of hidden layer output matrix.

Training and testing datasets are required to develop the ELM models. Training dataset is used to develop the ELM for prediction of liquefaction susceptibility of soil. For CPT data, 94 and 40 datasets have been used as training and testing datasets respectively. For SPT, 202 and 86 datasets have been used as training and testing datasets receptively. The program of ELM has been developed by using MATLAB.

RESULTS AND DISCUSSION

For CPT, radial basis function has been used as an activation function. Different number of hidden nodes has been tried to get best performance. Figure 1 shows the effect of number of hidden nodes on testing performance.

For MODEL I, the developed ELM gives best performance by using 12 hidden nodes (see Figure1). Training and testing performance have been determined by using the following equation.

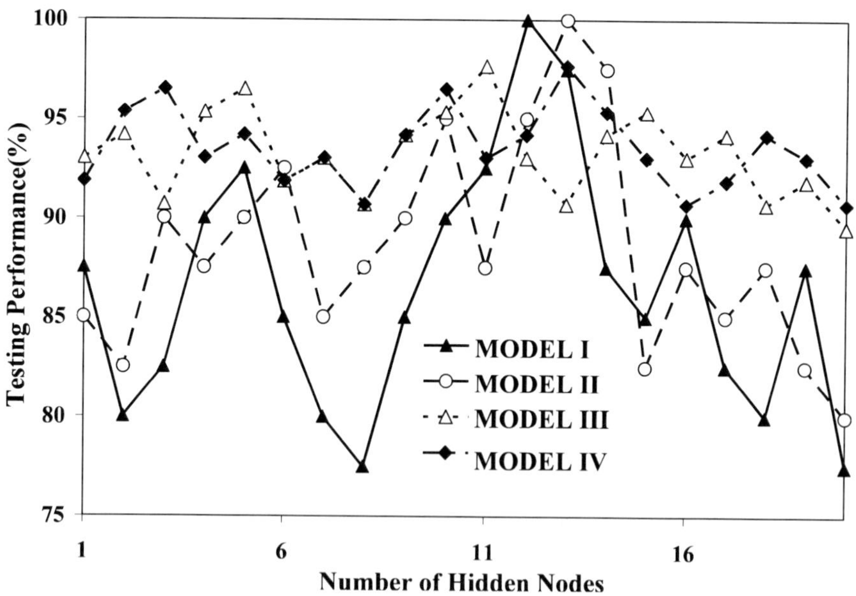

Figure 1. Effect of number of hidden nodes on testing performance for the different models.

$$\text{Testing / Training performance(\%)} = \left(\frac{\text{No of data predicted accurately by ELM}}{\text{Total data}} \right) \times 100 \tag{5}$$

The developed ELM gives 100% training as well as testing performance for MODEL I. For MODEL II, 13 hidden nodes give best performance. There is no misclassification in the MODEL II. Therefore, the performance of MODEL II is 100% for training as well as testing dataset. Figure 2 and 3 depicts the performance of MODEL I and MODEL II respectively.

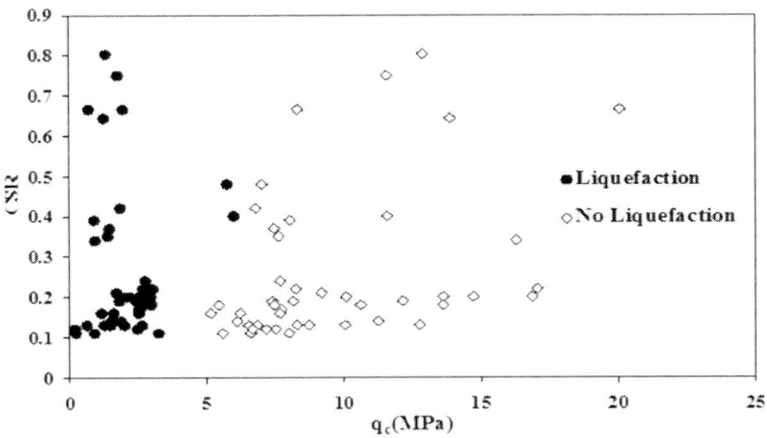

Figure 2. Plot between CSR and q_c for MODEL I.

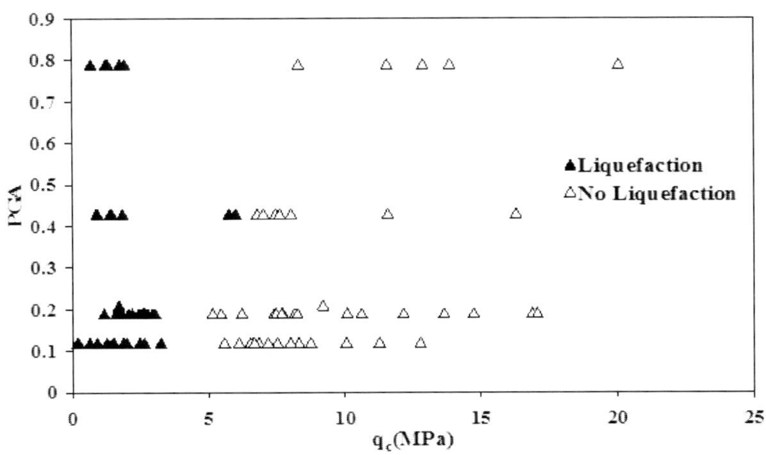

Figure 3. Plot between PGA and q_c for MODEL II.

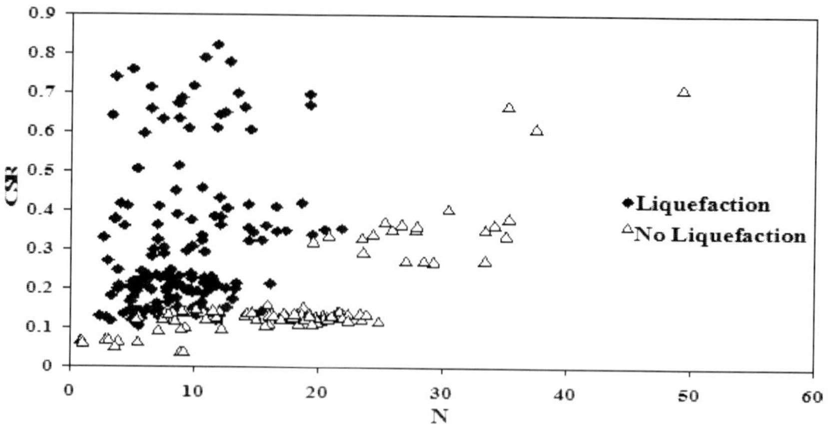

Figure 4. Plot between CSR and N for MODEL III.

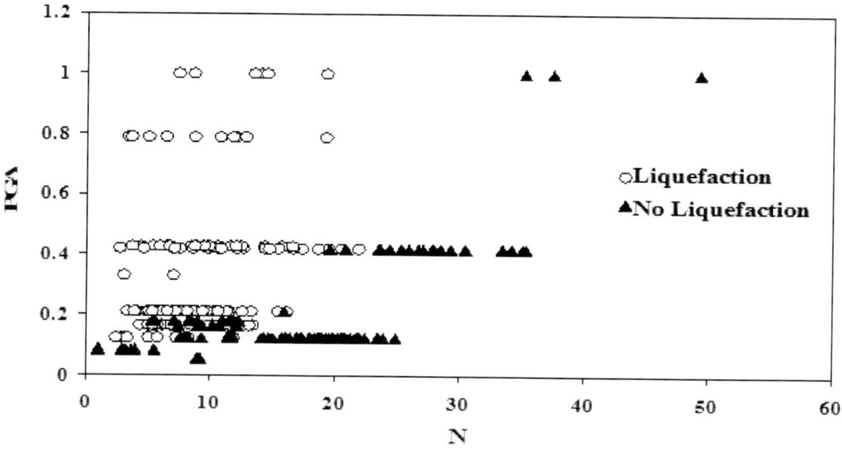

Figure 5. Plot between PGA and N for MODEL IV.

Radial basis function has been adopted as an activation function for developing SPT based model. For MODEL III, ELM gives best performance for 11 hidden nodes (see Figure 1). For training dataset, the developed ELM correctly classified 198 datasets out of 202. Therefore, the training performance is 98.01%. The performance of testing is 97.67% for MODEL III. MODEL IV gives best performance for 13 hidden nodes (see figure 1). The performance of MODEL III and IV is same. Figure 4 and 5 illustrates the performance of MODEL III and MODEL IV.

Table 2. Comparison between the developed models

MODEL	Training Performance (%)	Testing Performance (%)
I	100	100
II	100	100
III	98.01	97.95
IV	98.01	97.95

This article gives a comparative study between the developed four models. Table 2 shows the comparison. It is clear from Table 2 that CPT based model gives the better performance.

Another important observation from Table 2 that the performance of MODEL I and II is same. Same observation is applicable to the SPT based method also. So, the determination of CSR is not required to apply the developed models.

CONCLUSION

This article adopts ELM for determination of liquefaction susceptibility of soil based on SPT and CPT. Four models have been developed. CPT based models give better prediction than the SPT based models. The developed ELM shows that CSR is not required for determination of liquefaction susceptibility of soil. This study shows that q_c and PGA are sufficient parameters for prediction of liquefaction susceptibility of soil. The developed ELM gives an efficient and quick model for prediction of seismic liquefaction potential of soil.

REFERENCES

Ambraseys NN. (1988) Engineering seismology. *Earthquake Engineering and Structural Dynamics*. 17: 1–105.

Andrus, R, D, & Stokoe KH, Chung RM. (1999). *Draft guidelines for evaluating liquefaction resistance using shear wave velocity measurements and simplified procedure*: NISTIR 6277, National Institute of Standards and Technology. Gaithersburg.

Arango I. (1996). *Magnitude scaling factors for soil liquefaction evaluations.* Journal of Geotechnical Engineering, ASCE. 122(11): 929–936.

Horata P, Chiewchanwattana S, Sunat K. (2013). *Robust extreme learning machine. Neurocomputing*. 102: 31–44.

Huang GB, Zhu QY, Siew CK. (2006) *Extreme learning machine: theory and applications.* Neurocomputing. 70(1–3): 489–501.

Huang GB, Zhu QY, Siew CK. (2006). *Extreme learning machine: theory and applications.* Neurocomputing. 70(1–3): 489–501.

Hwang JH, Yang CW. (2001). *Verification of critical cyclic strength curve by Taiwan Chi-Chi earthquake data.* Soil Dynamics and Earthquake Engineering. 21: 237-257.

Ku CS, Lee DH, Wu JH. (2004). *Evaluation of soil liquefaction in the Chi-Chi Taiwan earthquake using CPT.* Soil Dynamics and Earthquake Engineering. 24: 659-673.

Seed HB, Idriss IM, Arango I. (1983). *Evaluation of liquefaction potential using field performance data.* Journal of Geotechnical Engineering Division, ASCE. 109(3): 458-482.

Seed HB, Idriss IM. (1967). *Analysis of soil liquefaction*: Niigata earthquake. J. Soil Mech. and Foundation Div, ASCE. 93(3): 83-108.

Seed HB, Idriss IM. (1971). *Simplified procedure for evaluating soil liquefaction potential.* Journal of Soil Mechanics and Foundation Division, ASCE. 97(9): 1249-1273.

Seed HB, Idriss IM. (1982). *Ground motions and soil liquefaction during earthquakes*: Earthquake Engineering Research Institute Monograph Oakland: California.

Seed HB, Tokimatsu K, Harder LF, Chung RM. (1984). *Influence of SPT procedures in soil liquefaction resistance evaluation.* Rep. No. UCB/EERC-84/15, Earthquake Engrg. Res. Ctr., Univ. of California. Berkeley: California.

Skempton AW. (1986). *Standard penetration test procedures and the effects in sands of overburden pressure, relative density, particle size, aging and overconsolidation*. Géotechnique. 36(3): 425-447.

Wang X, Chen A, Feng H. (2011). *Upper integral network with extreme learning mechanism. Neurocomputing.* 74(16): 2520–2525.

Yeu CWT, Lim MH, Huang GB, Agarwal A, Ong YS. (2006). *A new machine learning paradigm for terrain reconstruction.* IEEE Geosci Remote Sens Lett. 3(3): 382–386.

Youd TL, Idriss IM, Andrus RD, Arango I, Castro G, Christian JT, Dobry R, Finn WDL, Harder Jr. LF, Hynes ME, et al. (2001). *Liquefaction resistance of soils: summary report from the 1996 NCEER and 1998 NCEER/NSF workshops on evaluation of liquefaction resistance of soils.* Journal of Geotechnical and Geoenvironmental Engineering, ASCE. 127(10): 817-833.

Youd TL, Noble SK. (1997). *Magnitude scaling factors. Proc., NCEER Workshop on Evaluation of Liquefaction Resistance of Soils,* Nat. Ctr. for Earthquake Engrg. Res., State Univ. of New York at Buffalo. 149–165.

Zong W, Huang GB, Chen Y. (2013). *Weighted extreme learning machine for imbalance learning*. Neurocomputing. 101: 229–242.

In: Earthquakes
Editor: Wayne Coleman
ISBN: 978-1-53610-342-7

Chapter 3

MITIGATING GROUND SHAKING CONSTRUCTION ACTIVITIES IN COASTAL CITIES OF NIGERIA: AN EARTHQUAKE PREVENTIVE MEASURE

P. O. Awoyera[1], B. U. Ngene[1], G. A. Adeyemi[1] and P. A. Aderonmu[2]
[1]Department of Civil Engineering, Covenant University, Ota, Nigeria
[2]Department of Architecture, Covenant University, Ota, Nigeria

ABSTRACT

This chapter presents a review of earthquake phenomenon in Nigeria, the occurrence and available data. The review also includes recommendation to government agencies and designers of Engineering Structures on the need to accommodate mitigating efforts to avoid possible future disasters occasioned by earthquake occurrence. Nigeria may not be presently in high impact zone of earthquake phenomenon, but what the future portends cannot be easily predicted. Hence there is necessity for long term planning to avoid being caught unaware of the possible effect of sudden high impact earthquake in its shores. Thus, the need for the application of cutting edge Information Technology and adoption of mitigation techniques in the reduction of earthquake effects cannot be overemphasized. Environmental impact assessment prior to construction should indicate the seismic performance of concrete frames, perhaps under simulated earthquake excitations. Also significant is the effect of rapid urbanization and consequent construction of heavy engineering structures that require deep foundation in coastal cities (Lagos, Warri, Calabar and Port Harcourt) of Nigeria. The effect of such construction activities can trigger earth movement in the area. Thus, the destructive impacts of earthquake, both on lives and properties could be ameliorated by taking mitigation measures of risks connected to earthquake. Mainly, through adherence to effective town planning and land use plans, enforcement of standard building codes and good construction practices. Moreover, adequate early warning systems, awareness creation and community preparedness would serve as protective mechanism. Finally, hazard and risk mapping and analyses should be carried out on a national and regional basis, so as to enhance emergency preparedness, rescue operation planning and resources allocation.

Keywords: earthquake, construction, natural hazard, seismic analysis, earth movement

1. Introduction

Earthquake and other natural disasters are mostly categorized as natural phenomena. However, past occurrences have shown that some man-made activities on the subsoil layer of the earth could also trigger an earthquake (Kreimer and Munasinghe, 1991; Menshikov et al., 2011; Facciorusso et al., 2016). Activities that impair or cause severe vibrations in the subsoil, such as during a construction work for structural foundations or any underground structures could contribute to the occurrence of the aforementioned disasters. Often times, majority of people in developing countries tend to overlook the influence of deep construction activities as a source of earthquake. During construction of boreholes, tunnels, and deep structural foundations, there is vibrating response on the earth. Apparently this accumulates based on the number of such construction activities, and invariably increase the vibration in the ground. For coastal cities, which are mostly dominated by weak soil strata, the vibrating effect that result from construction works can constitute major cracking in the subsoil layers. Moreover, such area is prone to earth movement (Walker, 2014; Baeda et al., 2015); an event that is common occurrence prior to earthquake.

Nigeria has four (4) major coastal cities and they are heavily populated. In addition, the cities are also high business activity centers. Figure 1 shows

Nigeria coastal cities as marked on a google map. Hence, due to increasing population size, business and urbanization in these areas, construction of infrastructures are also on the increase. Howbeit, it is important to evaluate the impact of heavy constructions in coastal cities on earthquake occurrence, whilst also taking necessary precautions in the building construction approach.

In the same vein, professional practices in the built environment can help devalue earthquake impact on structures and the people. It is noteworthy that approach to building construction can reduce the level of destruction during earthquake (Awoyera et al., 2016). According to Dimova and Negro (2005), the quality of construction is one of the main factors that affect the seismic vulnerability of structures. Thus, earthquake effect on buildings can be minimized, with proper considerations right from the conceptual design to the construction stage.

From the past evidences of potential earthquakes experienced in Nigeria, the National Space Research and Development Agency (NARSDA) has articulated that the nation is not immune from earthquake occurrence. Therefore, it is pertinent that adequate measures be put in place by relevant government agency, in preparedness against any future hazard occurrence.

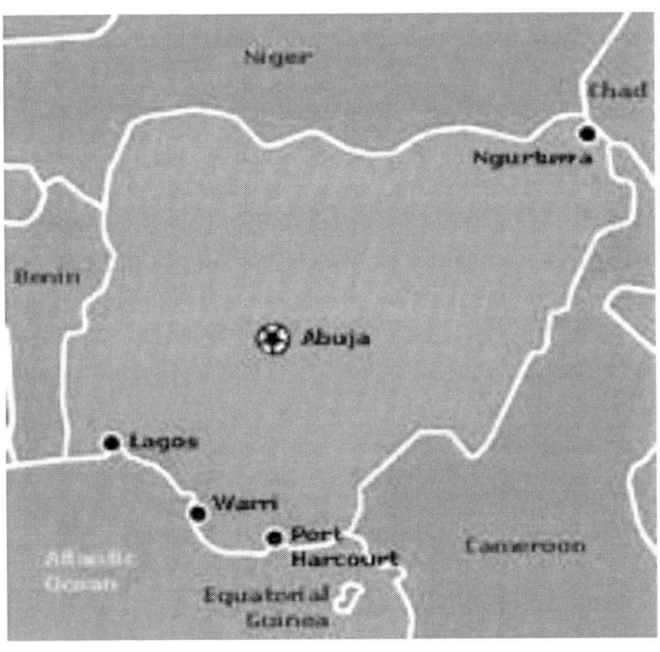

Figure 1. Coastal cities of Nigeria (Source: google maps).

It is well known that earthquakes are sudden event that usually claim large number of lives and properties. Unfortunately, this kind of hazard gives no one an impending notice.

As a result, this chapter is focused on evaluating ground shaking construction activities in the coastal cities of Nigeria and their potential influence on occurrence of earthquake. This study additionally renders recommendations to the Government agencies and professionals involved in the building and construction sector on measures that would control any imminent hazard occurrence.

2. Natural Hazards Occurrence in the World

A hazard may be man-made or due to natural process that may be harmful to a group of individuals or the environment that are in contact with it (Afegbua et al., 2011). Examples of hazards include earthquakes and earthquake induced tsunamis, volcanic eruptions, landslides, floods, hurricanes, and so on. Natural hazards are distributed in every part of the world and no country or continent can claim not to have its fair share of natural hazards. The only difference is in the intensity of one event in one location as against another.

The world at large has recorded numerous hazards, most of which have produced severe effect on human lives. The center for research on the epidemiology of disasters, in her 2013 report, indicated that natural disasters still killed a significant number, a total of 9,655 people (annual average 2002-2011:107,000) and 124.5 million people become victims worldwide (annual average 2002-2011:268 million). Moreover, economic disruption resulting from natural disasters did show an increase to above average levels (143 billion 2012 US $), with estimates placing the figure at US$ 157 billion.

Hazard effects can be described as primary, secondary and tertiary (Nelson, 2014). According to Nelson, p*rimary effects* are as a result of the process involved, such as collapse of buildings during an earthquake, landslide, or hurricane. *Secondary effects* are as a result of primary effect, such as, fires ignited as a result of earthquakes, disruption of electrical power and water service as a result of an earthquake, flood, or hurricane, or flooding caused by a landslide into a lake or river. Whereas, *tertiary effects* are long-term effects that are set off as a result of a primary event, such as, loss of habitat caused by a flood, permanent changes in the position of river channel caused by flood, crop failure caused by a volcanic eruption and nuclear hazard and their effect such as in the tsunami that occurred in Japan.

Studies have shown that less developed countries are more vulnerable to natural hazards than are industrialized countries (Benson and Clay, 2006), due to their level of understanding, education, infrastructure, building codes, and so on. Moreover, Nelson inferred that poverty plays a role in these calamities, in that poverty leads to poor building structure, increased population density, and lack of communication and infrastructure.

Consequently, a regular awareness programme is required to sensitize people in the developing countries about the danger of natural disaster. More also, there is need to curb man-made activities which could cause hazard to the environment.

2.1. Impacts of Hazards on Structures

Due to the devastating effect of earthquake load on structures and as a result of very low knowledge of their occurrence mechanism in most developing countries like Nigeria, Haiti etc., the need to study the phenomena of earthquake and other disasters become imperative. Most civil engineering structures are designed as being in a state of static equilibrium (Weekes, 2012). It is so because dynamic loads are treated as insignificant in magnitude and rate at which the load is applied. Major dynamic load effect that may be considered in design may include wind, ground motion (earthquake, underground explosions) and impact loading. An example of such impact loading was the impact of plane used during terror attack on World Trade Centre, USA on 11th September 2001. It is significant to note that Nigeria has not witnessed a major catastrophic effect of earthquake, being a non-earthquake prone zone of the world and having earthquakes of low intensity.

Major hazards and their impacts on structures are discussed as follows:

(i) Earthquake: Earthquakes are the result of displacement along geological faults and faults are fractures on the earth surface along which displacement has taken place (Weekes 2012). Since most faults are locked in, the forces causing displacement increases in magnitude overtime until the shear strength causing the obstruction is exceeded resulting in the release of significant amount of strain energy. It is the release of such energy that is said to cause earthquake according to the elastic rebound theory shown in Figure 2.

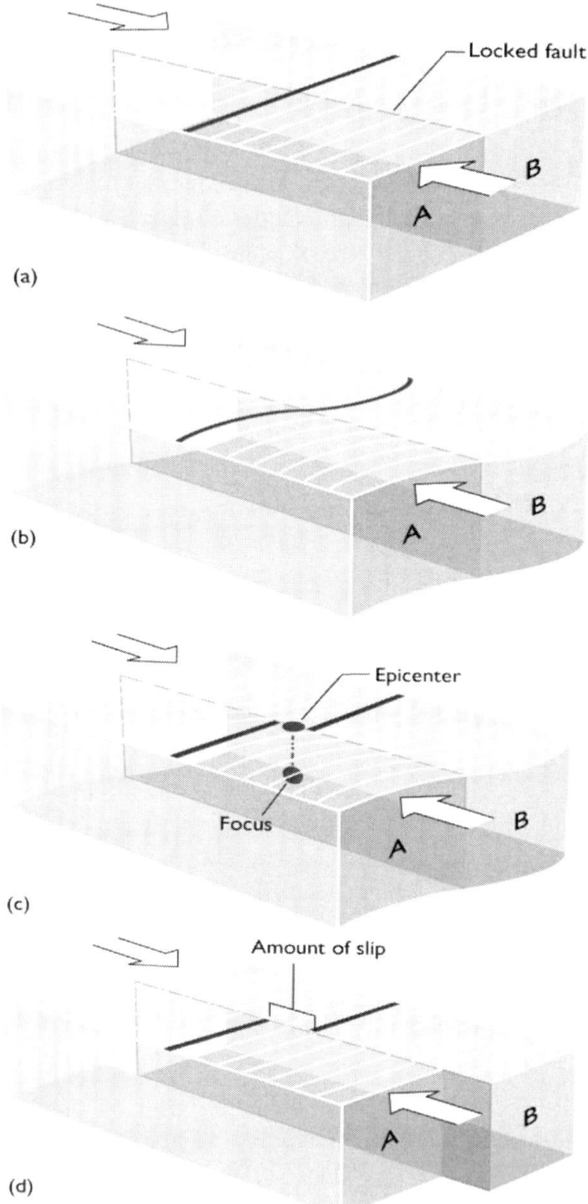

Figure 2. The elastic rebound theory of an earthquake (after Press & Siever, 2001).

Other causes of earthquakes are the movement of magma associated with volcanism, deep seated movements in the earth's crust. Earthquakes arising from these are however much smaller than those due to the elastic rebound theory.

Earthquake phenomenon is known to induce the following hazards: ground shaking, soil amplification, surface rupture, regional subsidence, densification, liquefaction, slope instability and tsunami and seiches.

(ii) Volcanoes: When two tectonic plates movement is towards each other and one plate overrides the other the overridden plate is thrust downward into the mantle below. This process is called subduction. The process of subduction produces deep-sea trenches, adjacent mountain ranges of folded and faulted rocks and magmatic belts. The magmatic belt can be a mountain range on land or a chain of volcanic islands, called an island arc, on the sea floor. As the subducted plate plunges into the underlying hot mantle the increase in temperature causes water and other volatile substances to "boil off". They join hot materials in the wedge of mantle above and this result in melting. This process is the source of magma that feeds the overlying chain of volcanoes. The zone of subduction is characterised by very large forces that result in the faulting that triggers shallow and deep focus earthquakes. Convergent or destructive plate boundaries are the home for the world most dangerous volcanoes and the largest and most devastating earthquakes.

(iii) Tsunamis: A tsunami is a series of large waves of extremely long wavelength and period usually generated by a violent, impulsive undersea disturbance or activity near the coast or in the ocean. Large tsunami waves are developed when there is a sudden displacement of a large volume of water due to a meteorite impact or a large coastal or submarine landslide, or if the sea floor is suddenly raised or dropped by an earthquake. The widely reported East Japanese tsunami of 11th March, 2011 is a good example of how devastating the effect of a tsunami can be. The waves travel out of the area of origin and can be extremely dangerous and damaging when they reach the shore. Part of the damages caused by the Tsunami is the melt down of the Fukushima Nuclear power plant also, part of the debris from the damages from the magnitude-9 earthquake induced Tsunami is being washed up the coast of North America two year after the incident Siringoringo and Fujino (2015), and Fritz et al. (2012). Tsunami is Japanese word that, when translated into English, means "harbour wave".

A Seiche is similar to a tsunami, except that it develops in inland waters, such as large lakes. The tsunami waves are completely unrelated to the astronomical tides, which are caused by the extra-terrestrial, gravitational influences of the moon, sun, and the planets.

2.1.1. How They (Earthquakes, Volcanoes and Tsunamis) Develop

The developmental processes for the three phenomena are similar as the movement in the earth crust seems to be the trigger for their occurrence. The plot of past earthquake occurrences indicates that it is prevalent along the sinuous lines across the Earth's surface which is explained by the theory of plate tectonics. According to this theory, the earth's surface is split up into a series of plates, known as lithospheric plates and there exist an association between areas of high seismicity and plate boundaries. Figure 3 shows the various demarcations of world plates that make up the theory of tectonic plate.

Below the plates is the upper mantle known as the asthenosphere. It is the asthenosphere semi-molten material that is moving by convection. The plates move relative to one another on average by 6cm/yr and it is thought that this movement is essentially driven by the convection cells within the asthenosphere. The nature of movement determines the type of plate boundary which also determines the nature of earthquake formation classified into divergent, convergent and transformation boundaries (Weekes 2012).

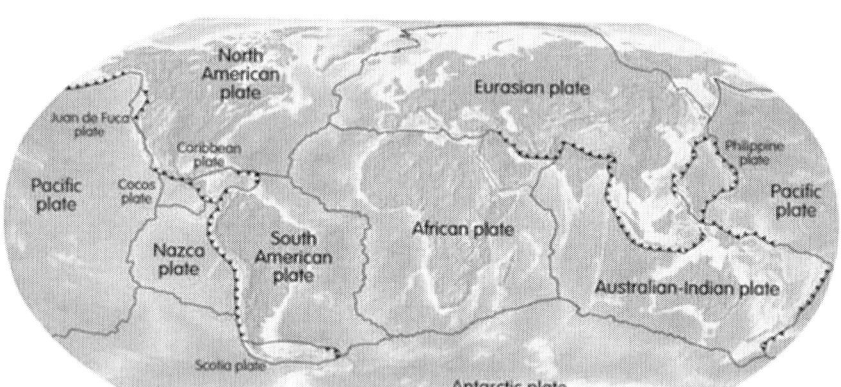

Figure 3. Various demarcations of world plates that make up the theory of tectonic plate (Source: Alvin Lucero, 2016).

Table 1. Approximate correlation between Local Magnitude ML and ground motion, seismic energy yield, peak ground acceleration a_{max} and duration of ground shaking (In Weekes (2012) as Sourced from Yeats et al. (1997), Gere and Shah (1984) and Housner (1970))

Local Magnitude ML	ML Ground Motion 100km from epicentre	TNT for Seismic energy yield at the focus of the earthquake.	Typical peak ground acceleration a_{max} near the vicinity of the fault rupture.	Typical duration of ground shaking near the vicinity of the fault rupture.	Comments
1.	0.0048μm	13.6kg			Equivalent to a large blast on a construction site. Generally not felt, but recorded.
2.	0.048μm	1 Tonne			Equivalent to a large quarry or mine blast. Generally not felt, but recorded.
3.	0.48μm	26 Tonnes			Smallest earthquake commonly felt.
4.	4.8μm	900 Tonnes			Equivalent to a small nuclear weapon. Often felt, but rarely causes damage.
5.	0.048mm	29 kTonnes	0.09g	2s	slight damage to well-designed buildings. Can cause major damage to poorly constructed buildings over small regions.
6.	0.48mm	900kTonnes	0.22g	12s	Can be destructive in areas up to about 100km across where people live.
7.	4.8mm	29MTonnes	0.37g	24s	Equivalent to largest nuclear weapon. Can cause serious damage over larger areas (e.g., Hyogo-Ken Nanbu, Japan quake, 1995)
8.	48mm	900Mtonnes	0.5g	34s	Great earthquake. Can cause serious damage in areas several hundred kilometres across (e.g., San Francisco quake, 1906)
9.	480mm	29 GTonnes			Massive earthquake (rare). Complete destruction over wide area (e.g., Chilean quake, 1960, Sumatra quake, 2004)

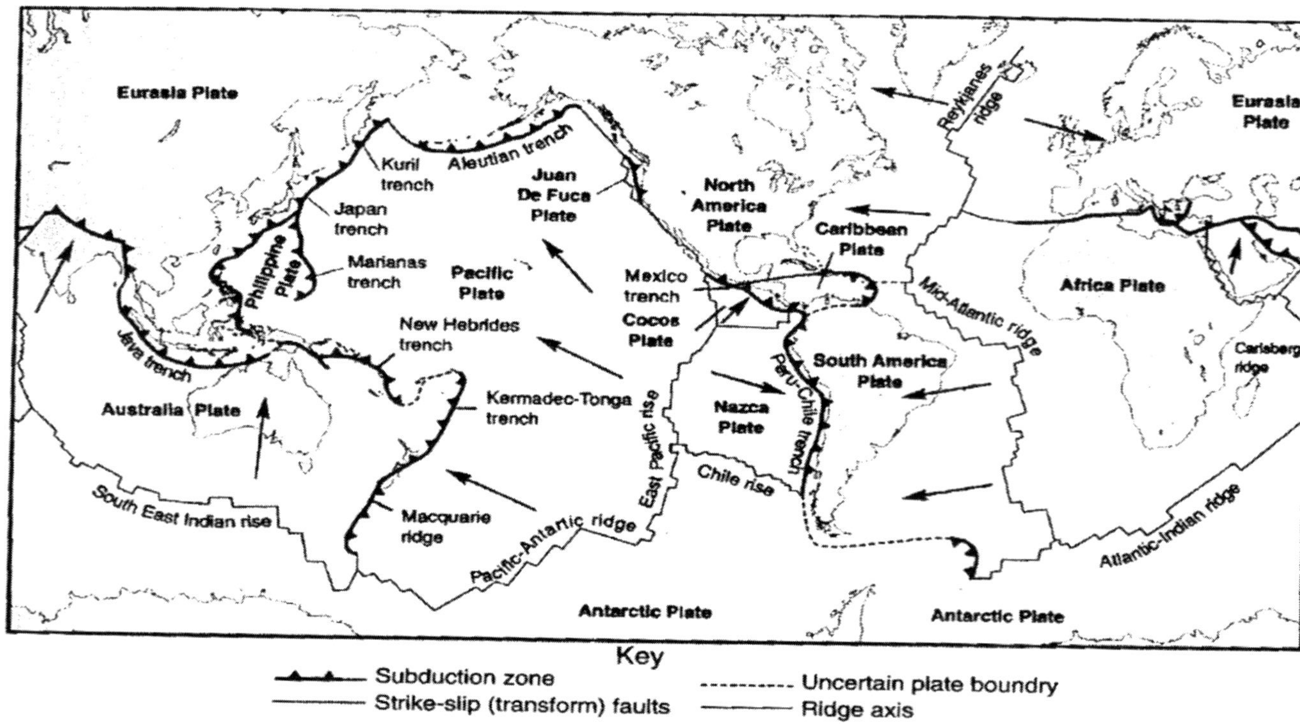

Figure 4. The major tectonic plates of the Earth. Arrows indicate directions of plate movement (from Kramer, 1996, source: Weekes 2012.

The unit of measure of earthquake is the Richter scale also known as the Local Magnitude scale which is a logarithmic scale. Earthquake effects are rated using the Modified Mercalli (MM) intensity scale, which ranges from I (imperceptible) up to XII (total destruction). From Table 1 it is shown that the earthquake wave displacement amplitude, that is, column 2 increases by a factor of 10 for every 1 unit increase of the local magnitude ML.

2.1.2. Global Distribution of Earthquakes, Volcanoes and Tsunamis

According to the theory of tectonic plate, most earthquakes occur around the plate boundaries. This explains the prevalence of this phenomenon at some locations in the world and not in others as shown in Figure 4.

Global Distribution of Earthquake

Earthquake phenomenon is known to have occurred around the world in a random distribution, but following the location of tectonic plates. Some examples of earthquake occurrence around the world will include the following:

The 1978 earthquake in Iran recorded a 7.4-magnitude quake that is said to have killed over 15000 people. In India State of Gujarat a 7.7-magnitude earthquake flattened towns and villages in Kutch region in January 2001.

It has been observed that part of the challenges of mitigating earthquake problems is the issue of access to sites such as those of Parkistan-administered Kashmir which occurred in October 2005. The earthquake in Haiti occurred on January 2010 with high devastation in terms of magnitude and level of destruction to a very poor nation in the Americas Figure 5. Huge numbers of human lives were lost some figures indicating as much as 200000 people died and also both government and private properties were lost in the 7.0-magnitude earthquake.

One of the worst cases of earthquake induced tsunami in recent history is the great east Japan Honshu coast earthquake of 11th March 2011. Though the country has recorded quite a number of major earthquakes in time past, this magnitude 9.0 earthquake according to the United Nations Environment Programme (UNEP), has been recognized as the most powerful earthquake in the country's history and one of the world's five strongest earthquakes since modern record keeping began in 1900, (Tanikawa, Managi, and Lwin, 2014).

The United States has a history of earthquake occurrences with California State among the highest in terms of number of occurrence and magnitude of each case. According to USGS (2016) the first strong earthquake listed in earthquake annals for California occurred in the Los Angeles region in 1769 and there have been several earthquakes thereafter with high level of loss of lives and properties. Figure 6 below shows earthquake distribution around the world.

Figure 5. Destruction of lives and properties were recorded in Haiti earthquake of 2010 (Source: BBC news).

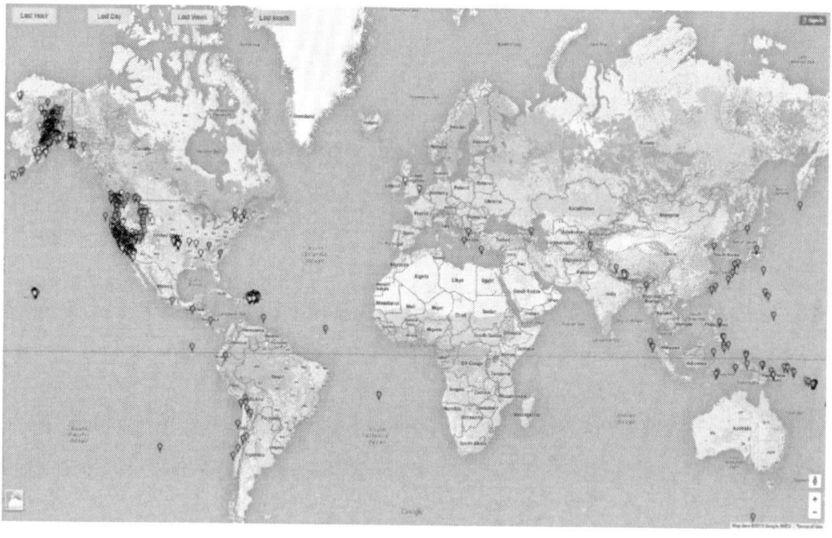

Figure 6. Earthquake distribution around the world with Nigeria and Africa relatively not represented.

From the Figure 6, it can be seen that large number of the earthquakes are concentrated in the western coastal part of the country where States of California and Alaska are located.

Currently it is being suggested that there is a link between sun magnetic fields and earthquake occurrence, Davidson (2016), however, research on this subject is still in its infancy.

Earthquakes in Africa

The continent of Africa may not currently be described as high earthquake zone, yet it has a share of the phenomenon especially in the South, Central and eastern region as indicated in Figure 7.

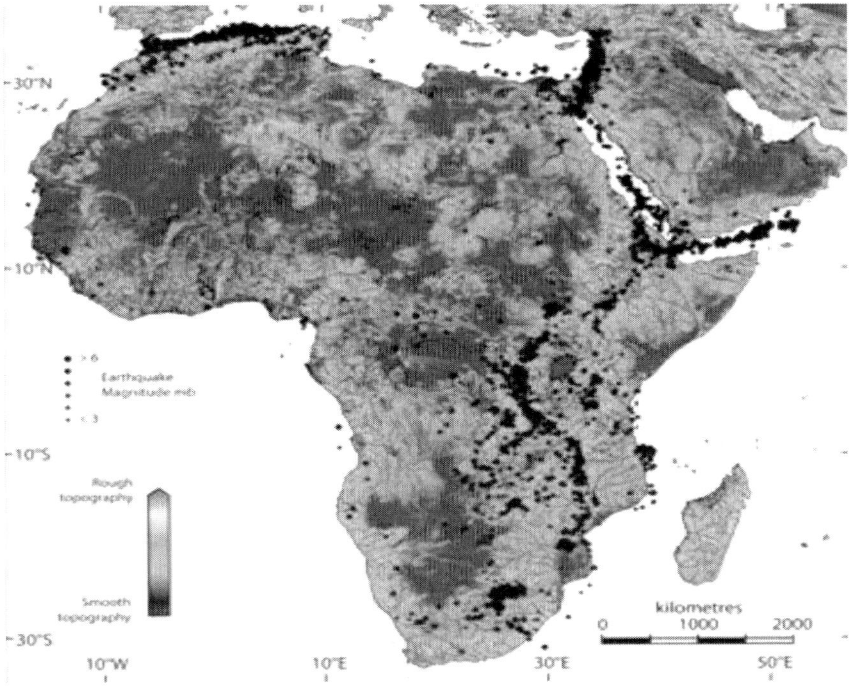

Figure 7. Earthquake Map of Africa showing areas of high concentration in the south and central Africa with active rift valley region also recording good number (Source: Geoff et al., 2012).

From this Figure 7, it is seen that South Africa, Central, East (the rift valley area) and parts of the North Africa bordering the Mediterranean Sea has experienced a great number of earthquakes. The West African zone where

Nigeria is located has relatively low number of earthquake occurrence. Major earthquakes in Africa are known to follow the divergent margin such as the rift system. The map of major earthquakes in Africa show that in 1995 there was a 7.2-magnitude earthquake in Sinai, Egypt and in 2006 there was a 7.0-magnitude earthquake in Mozambique (Africathistime, 2015). Both earthquakes are in the Great Rift Valley system.

Aside these, earthquakes occur also in the north-eastern part of the African continent around the Atlas Mountains which is related to the convergent boundaries between Africa and Eurasian plates Figure 3. According to Osagie (2008), it is wildly believed that African plate has not moved in the past 200 years which account for the low level of seismic activities in the continent.

Nigeria Earthquakes Phenomena

Lack of information and data on past and current earthquake occurrence in Nigeria has been the bane of proper analysis and design of structures to resist earthquake waves in the country. Research efforts such as this are therefore meant to update available information on such phenomenon, with the aim of making design Engineers, government agencies and private individuals and organizations aware of the issues involved.

Many earthquakes or earth tremors have been recorded in Nigeria with their epicentre in the country or in the neighbouring countries such as Benin Republic and Cameroon. Earth tremor is defined as micro-earthquakes of magnitude less than three (3) which is mainly movement of earth due to human activities such as in mining. Table 2 shows information on earthquake occurrence available mainly through oral historical records of occurrence of earthquakes in the Nigeria since record began in the early 20th century with the various locations involved.

A look at Table 2 and Figure 8 will indicate a high concentration of Nigeria earth tremors along the Ifewara-Zungeru fault line which is known to have a link with the Atlantic fracture system (Anifowose et al., 2010; Adepelumi et al., 2008; Olujide and Udoh, 1989: Olurufemi et al., 1986). Record of earth tremors in the eastern part of the country is acknowledged to be due to the presence of volcanoes around Lake Nyos in the Republic of Cameroon. Most locations as shown in the table does not have records of magnitude of earthquake occurrence because of lack of good seismographic network of equipment for measuring the earthquake phenomena, proper record keeping, that is, data storage and retrieval system in the country. The Nigeria Centre for Geodesy and Geodynamics in Toro, Bauchi State as recently established is to bridge the gap in terms of information generation and dissemination. When it is fully operational with optimum number of seismographic stations it will be expected to closely monitor earthquake activities in the country and provide early warning information that will help in managing this natural and sometimes man-made disaster.

Table 2. Earthquake occurrence in Nigeria with location and date of occurrence (Source: Osagie 2008; Akpan and Yakubu, 2010, Tsalha et al., 2015)

S/N	Location	Date	Intensity/ Magnitude	Remarks
1.	Warri (Delta State)	1923		
2.	Ohafia (Abia State)	1933		
3.	Lagos, Ibadan, Ile-Ife	22nd June 1939	6.5(ML)	
4.	Ibadan (Oyo State)	28th Jul 1948		
5.	Ohafia (Abia State)	2nd Jul 1961		
6.	Ijebu-Ode (Ogun State)	21st Dec,1963	V	
7.	Kundunu	23rd April 1981	III	
8.	Jalingo and Gembu	16th Oct.1982	III	Close to Cameroon Volcanic line
9.	Ijebu Remo (Ogun State)	12th Jul 1984	IV	
10.	Ibadan, Shagamu, Ijebu-Ode, Abeokuta	28th Jul 1984	VI	
11.	Ijebu-ode, Abeokuta (Ogun State)	2nd Aug, 1984	V	2s Earthquake
12.	Yola (Adamawa State)	8th Dec, 1984	III	Close to Cameroon Volcanic line
13.	Kombani Yaya	18th June 1985	IV	
14.	Obi (Benue State)	15th Jul 1986	III	
15.	Abeokuta (Ogun State)	1986		Underground water exploitation thought to be the cause
16.	Gembu (Taraba State)	27th Jan 1987	V	Close to Cameroon Volcanic line

Table 2. (Continued)

S/N	Location	Date	Intensity/ Magnitude	Remarks
17.	Akko	19th Mar 1987	IV	
18.	Kurba	24th May 1987	III	
19.	Amauzu Ede-Obala (Enugu State)	April 1988		12km land with surface cracks
20.	Lagos (Lagos State)	14th May 1988	V	
21.	Osererun Hills (Gombe State)	Sept 1988		
22.	Jere (Kaduna State)	5th April 1990	V	
23.	Ibadan (Oyo state)	27th June 1990	3.4(ML) magnitude	Micro-quake covers 16km radius
24.	Ijebu-Ode (Ogun State)	7th Nov1994	4.2(ML)	
25.	Okitipupa (Ondo State)	1997	IV	Close to Okitipupa Ridge
26.	Jushi-Kwari	15th Aug 2000	III	
27.	Benin (Edo State)	13th Mar 2000	IV	
28.	Ibadan, Akure, Ijebu-Ode, Abeokuta,Oyo	7th Mar 2000	4.7(ML)	
29.	Akure	7th May 2000	IV	
30.	Lagos	19th May 2001	IV	
31.	Lagos	8th Aug 2002	IV	
32.	Yola	Mar 2005	III	
33.	Lupma	25th Mar 2006	III	
34.	Rivers State	April 2006		
35.	Abomey-Calavi	11th Sept 2009	II	Close to Benin
36.	Abeokuta (Ogun State)	5th Nov 2011	4.4(ML)	
37.	Saki-Oke Ogun (Oyo State)	May 2016		

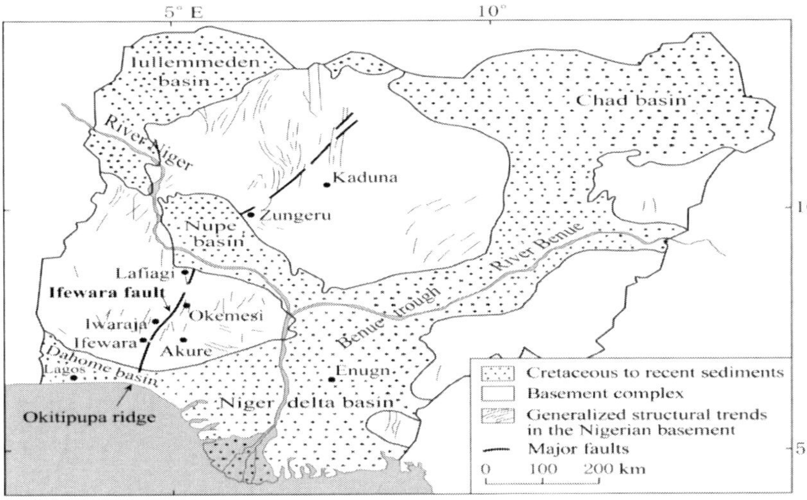

Figure 8. Map of Nigeria showing the Zungeru-Ifewara fault (Akpan and Yakubu, 2010) sourced from Afegbua (2011).

Though the occurrence earthquakes are known be of random distribution, the number of earthquake events that has occurred in Nigeria since records began is seen in Figure 9 below to be of negative skew. This is an indication, as per Adepelumi, 2009 prediction of increased earthquake events from 2008 that the country is going to expect more occurrences in the foreseeable future in terms of intensity, magnitude and numbers bearing in mind increasing number of earth shaking activities in Nigeria.

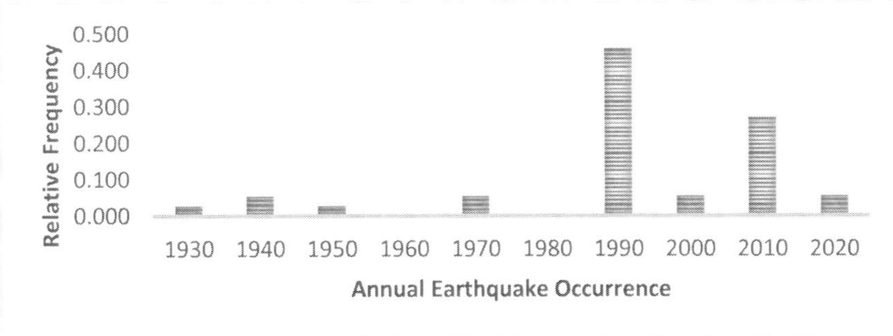

Figure 9. Distribution of annual earthquake occurrence in Nigeria (1923-2016) showing a negative skew.

2.2. Impact of Earthquake Phenomenon on Human

Every activity of human on earth has great impact on the sustainable presence of life on the surface of the earth. When high rise structures are erected either for housing or office accommodation, the construction of such structure with pile foundation may cause earth movements with consequent movement on fault lines in the earth crust in form of earthquakes or Tremors.

Earthquake phenomenon when they occur has devastating consequences on those living in the area. In Nigeria, though no major earthquake has taken place, the minor earth tremors recorded has caused untold hardship to the affected communities. The impact of such tremor can be summarized as follows:

(i) The recent earth tremor in Oyo state caused the people to live in constant fear of possible damage to life and structures.
(ii) The communities in the area started moving out of their ancestral home thereby causing internal displacement.
(iii) Disruption of people's means of livelihood may bring about greater hunger and poverty in the land.
(iv) The affected people are unsettled and confused especially the women and children.
(v) When high impact earthquakes are involved they could cause heavy damage to properties with possible loss of life.
(vi) Future economic life of the area may be affected as investors will be afraid to own or build structures in the area.

3. Ground Shaking Activities in Coastal Areas of Nigeria

The length of the Nigeria Coastline measures about 853km facing the Atlantic Ocean. The Niger Delta and Coastal areas of the Country represent the active sedimentation zone and yet because of the economic importance of the area to the nation, it is an area with high activities that causes earthquake. The following are some ground shaking activities that promote the propagation of earthquakes:

(i) Piling for high rise structures foundation: Foundation for high rise structures is located in firm soils of good bearing capacity to ensure effective transfer of load from the structure to the soil. Piling operation, however create waves within the soil structure that has the effect of causing densification and increase the stress build up that may trigger earthquake occurrence especially in the coastal areas. Soil liquefaction is one of the major factors that are known to affect the behaviour of piles founded in seismically active areas.
(ii) Blasting for oil exploration: Soil sounding for oil exploration involves sending waves at appropriate determined depth and spacing through which the quantity of oil available in an area can be obtained. Nigeria Coastal area is reach in oil and such activity takes place often. The environmental effect of such activity is that the blasts may increase the chance of reactivation of pre-existing fault or weak zone.
(iii) High speed drilling for the extraction of oil: Example is in the drilling for shale oil for which environmentalist question the safety of the location, suggesting that it causes sedimentation which may lead to earthquake occurrence.
(iv) Military activities especially during wars (Example Nigeria Civil War): Operations during wars have great impact in several forms. At such time heavy military equipment are moved from place to place and their resultant vibration send shock waves to the earth crust thereby increasing the strain energy which may result in earthquake especially in areas prone to it. Also when nuclear bombs are used as in the case of Japan during the Second World War their impact may trigger seismic activity. Though Nigeria has had Civil War in time past, the technology of today will pale to insignificance when compared to what obtains in the period 1966-1970.
(v) Extensive extraction of water resources through uncontrolled boreholes: Rate of exploitation of ground water resources has effect on earth movement as the voids created may cause subsidence which may result in built up of stresses that will trigger earth tremors. This is an area government agencies in Nigeria, especially those in the coastal areas will have to be proactively involved to check uncoordinated extraction of ground water.
(vi) Blasting of rocks for quarry mining and other mining activities: Industries such Cement factories, Coal mining, Rock quarries use a lot of explosives as access their raw materials. The blasts from such

industries have effect especially in areas close to existing fault line. The exploitation and use of such materials should be highly regulated for the overall safety of the inhabitants.

Most of these activities are taking place or have taken place in Nigeria coastal areas which is known to be of continuous sedimentary formation.

4. Crisis of Management of Earthquake Using Geographic Information System

The need for the application of cutting edge Information Technology and adoption of mitigation techniques in the reduction of earthquake effects cannot be overemphasized. Most destructive impacts of earthquake both on lives and properties could be ameliorated by taking mitigation measures of risks connected to earthquake. For instance, when there is adequate adherence to effective town planning and land use plans, enforcement of standard building codes and good construction practices, adequate early warning systems, awareness creation and community preparedness would serve as protective mechanism. As an earthquake preventive measure, hazard and risk mapping and analyses should be carried out on a national and regional basis to enhance emergency preparedness, rescue operation planning and resources allocation. Geographic Information System (GIS) has been applied in planning, emergency preparedness and recovery after an earthquake. GIS application formed an essential tool especially in Haiti after the 2010 devastating earthquake. By definition, GIS is an "organized collection of computer hardware, software, geographic data and personnel designed to efficiently capture, store, update, manipulate, analyze and display all forms of geographically referenced information". Sun et al. (2008) studied GIS-based information systems and its use in forecast and plan for natural hazards such as landslides or earthquakes. Particularly, GIS helps to represent geophysical data in their geographic location and enhance predictive capability. More also, analytical and mapping capacity of GIS could be applied in the production of seismic micro and macro zoning maps. Observations of destructive earthquakes have demonstrated that despite the same epicentral distances, damage is often more severe over soft soils than over firm soils or rocks (sun et al., 2008). Careful study of soil map couple with accurate and adequate soil analysis would enhance what to build where and how. The use of GIS can enhance overlay of different map (layers), such as population, soil, infrastructures and faults layer which can be viewed simultaneously, and it could be utilized in hazard zoning for emergency planning and preparedness. Risk assessment and hazard locations in relation to populations, property, and natural resources become easier applying GIS in vulnerability assessment.

In Nigeria situation, the level of awareness about GIS applications is growing. Parts of the constraints to GIS development in developing countries which Nigeria is not exempted have been identified to be 'lack of enough appreciation of the technology and limited understanding of GIS principles and associated methodology (Onosemuode and Dare, 2010). Also, lack of current, usable and reliable spatial data to make informed decision is one of the obstacles.

4.1. Earthquake Mitigation Process (Engineering Management)

Engineering Management cannot be entirely divorced from other forms of Management function which is summarized into the following: Planning, organizing, staffing, directing, and controlling. For the Engineer, however, proper design and construction function is very important in resolving most of the challenges faced in the field in addition to other management functions. Figure 10 show how earthquake occurrence can affect improperly designed and constructed shear wall.

Figure 10. Source: www.surrey.ac.uk Earthquake Engineering Module, 2012.

4.1.1. Structural design and Construction Implication

Effort to make a building resistant to earthquake can be envisioned right from the planning state to the actual construction. Despite the fact that most building owners are interested in the aesthetic look of a building, the architect has to contribute in the design to make the building resistant to earthquake. The indispensable role of the architect who initiates the building design and determines a number of issues relating to its configuration that have a major influence on the building's seismic performance cannot be overemphasized.

It has been articulated that architectural designs with typical shapes like L, T, X or H, which allow for more windows, are preferable in earthquake prone region (Guevara et al., 1992).

It is noteworthy that the structural layout of a building in a seismic region is of substantial importance in the ability of the structure to resist earthquakes.

At the design state, thus in line with consideration of (BS EN1998 and Uniform Building Code), uniformity in the development of the structure along the height of the building is required, since it tends to eliminate the occurrence of sensitive zones where concentrations of stress or large ductility demands might prematurely cause collapse. Similarly, Dimova and Negro (2005) emphasized the need for building designs to be based on requirements of modern codes for seismic design, which rely on high behaviour factor supplies, the assessment of the effects of poor quality of execution in otherwise well-conceived and well-designed structures becomes an important problem. Moreover, the knowledge of the epicentre and hypocentre is required to be able to determine the appropriate foundation for the structure otherwise the problem of stability will arise (Vere-Jones, 1995).

Construction of buildings in earthquake region requires adequate quality control from design through material selection and construction. Among other materials; structural steel, due to its large ductility and high strength-to-weight ratio is an ideal material for earthquake resistance in buildings (Duggal, 2007). From observation of damage to modern buildings after almost all recent earthquakes, Dimova and Negro (2005) reported cases of poor quality of materials, inadequate detailing of reinforcement and absence of capacity design principles. Figure 11 below is an example of pounding of column base during earthquake phenomenon.

Figure 11. Source: www.surrey.ac.uk Earthquake Engineering Module, 2012.

4.1.2. Risk Mitigation and Management

A fault has been defined by Rickett, Loftin and Merritt (2004) as zones of crushed rock between two tectonic plates in the earth crust. Such movement in the earth crust travel at high speed and on the earth surface thereby causing damages to lives and properties on their part. To mitigate the risk of such disaster the following are required.

(i) Identify the type of structures prone to damage and hence possible collapse such as non- ductile buildings. Similar to that is to identify infrastructures such as buildings, bridges, oil and gas infrastructures etc constructed across surface fault lines.

(ii) Civil and Structural Engineers in Nigeria Coastal area should be involved in identifying such failure-prone structures and take necessary actions to mitigate the risk.

(iii) For the two items above to be effective, the Parliaments in the States involved will need to enact laws to that will help to mitigate the hazards that may affect human occupancy in the area or location. The laws are meant to prevent

(iv) For existing structures, mitigation may include relocation of citizens from identified fault lines. Such structure as may be located in an area showing soil liquefaction or possible seismically induced landslide should be dismantled and the land converted for other purposes such as agriculture.

(v) Existing hospitals, schools, fire stations etc should be replaced or strengthened to obtain structures with earthquake resistance.

(vi) There should be conscious efforts to measure and map Nigeria earthquake occurrence and classify same to enable proper monitoring.

(vii) Nigeria Emergency Management Agency should have an arm with upto date information on possible actions to be taken during earthquake occurrence. The case of an earth tremor that occurred in Saki in Oke Ogun area of Oyo State, Nigeria as reported in Vanguard Nigeria Online (2016) illustrates the point of coordination very succinctly. According to the report and subsequent News items on the issue, it took the State government over a month to visit the community impacted and the high powered team only came to assure the people that they are taking action to determine the cause of the tremor and possibly solve the problem.

(viii) The current state of living in denial of earthquake occurrence in Nigeria should be countered with adequate information by responsible agencies and organization such as Nigeria Engineering Professional bodies.

(ix) Proper Engineering design of structures to take care of earthquake possibilities is advocated. Zahi Baroudi (2016) has identified the use of the following methods in designing building structures to resist earthquakes.

- By use of Base isolation method which is the making of the structure free from movement during earthquake excitation by the use of frictionless rollers.
- The use of seismic dampers as against structural elements, example, diagonal braces which absorb part of the energy transmitted through them. Types of dampers may include viscous, friction and yielding damper.

(x) Nigeria in 2006 established the Centre for Geodesy and Geodynamics in Toro to manage the Network of Seismographic stations, Tsalha (2015). Currently the Centre has established six stations to monitor earthquake activities with non in coastal cities of the country. It is therefore being advocated that more stations be opened for continuous monitoring and early warning especially in the coastal area where lots of ground shaking activities take place.

4.1.3. Future Expectation of Earthquakes in Nigeria

Adepelumi (2009) has suggested an increase in number and intensity of earthquake events in Nigeria especially in Ijebu-Ode axis with m $\geq$ 5magnitude between 2008 and 2038. Based on this statistical prediction, it is necessary for Structural Engineers and Planners to be conscious of this and therefore rethink the issue of earthquake effect more seriously.

At least three earthquake events have been recorded in the country since 2009, Table 2, which suggest the need to start action with minimum of properly equipping the earthquake Centre's that will be able to record such events. The unorthodox method of oral history should be replaced with up to date devices that can record both earth tremors and earthquakes of high intensity.

4.1.4. Professional Suggestions and Pedagogic Proposals

- Design Professionals need to incorporate in their digital works the seismometry and computer processing time of recorded data, in order to produce near-real time analysis products important for post-earthquake emergency response (Gee et al., 2003; Dreger and Kaverina, 2005; Wald et al., 1999).
- Architects and Engineers also need to synergize the benefits of their professional 3D softwares (AUTOCAD, BIM, Revit etc.) with realistic 3D numerical models in order to solve the problem of the sequential development of the wave mixtures as the waves pass through different geological structures. It creates room for probabilistic estimation of basic engineering decisions.
- Through the support of National University Commission, Council of Registered Engineers(COREN), Architect Registration Council of Nigeria and other Accreditation gate keepers; curriculum of design course works need to include ingredients of seismic knowledge and pragmatics into the schools' curriculum.
- Continuous Professional Capacity Development Workshop, seminars and conferences need to be emphasized in activities of the field practitioners.
- Problem-based Learning Model with real life experience needs to be included as part of the curriculum; where students' work assessment during excursions, field trips and students' industrial work experience.

Conclusion

Osagie (2008) has identified seismically active zones in Nigeria to include South-west zone, the Cross River Basin, Gongola trough around the current Adamawa/Taraba States and the Benue valley.

Nigeria Earthquake distribution should be mapped into zones, according to their peak ground acceleration corresponding to the seismic zone factor. This is necessary so as to enable design engineers make use of such resources in design decision making.

Every construction activity especially in the coastal areas must have appropriate environmental impact assessment, which should include method programme to reduce ground shaking activities in the zone.

Also, adoption of Geographic Information Systems and Remote Sensing is necessary to enhance emergency preparedness, mitigate the negative impacts of earthquakes on human lives and property. Availability of current, reliable and usable spatial and attribute data should be made a priority by the public institutions responsible for this to enhance adequate planning and institutional development.

Acknowledgment

The authors appreciate the effort of the NOVA editorial team for making comments that significantly improved the quality of this chapter.

References

Adepelumi A. A. (2009). *Short-term probabilistic forecasting of earthquakes occurrence in South-Western Nigeria*, Technical Report Submitted to the Centre for Geodesy and Geodynamics, Toro, Nigeria.

Adepelumi, A.A., Ako, B.D., Ajayi, T.R, Olorunfemi, A.O., Awoyemi, M.O., and Falebita, D.E (2008). Integrated geophysical studies of the Ifewara transcurrent fault system, Nigeria. *J. Afr. Earth Sci.,* 52: 161-166.

Afegbua, K. U., Abubakar, Y. T., Umo, A. O., Dauda, D. and Saturday, U. E. (2011). Towards an integrated seismic hazard monitoring in Nigeria using geophysical and geodetic techniques. *International Journal of the Physical Sciences,* 6(28), pp. 6385-6393.

Akpan O.U. and Yakubu T.A., (2010). A review of earthquake occurrences and observations in Nigeria. *Earthq. Sci.,* 23(3), 289-294.

Anifowose A.Y.B., Oladapo M.I., Akpan O.U., Ologun C.O., Adeoye-Oladapo, O.O., Tsebeje S.Y., and Yabuku T.A (2010). Systematic multi-technique mapping of the southern flank of Iwaraja fault, *Nigeria. J. Appl. Sci. Tech.,* 15(1-2): 70–76.

Awoyera P. O., Ogundeji J. F., and Aderonmu, P. A. (2016). Simulated Combined Earthquake and Dead Load Lateral Resistance Building Systems using Nigeria Seismic Data. *Journal of Materials and Environmental Science* 7 (3) (2016) 781-789.

Baeda, A. Y., Rachman, T., Umar, H., Suriamihardja, D. A. (2015). Mitigation Plan for Future Tsunami of Seruni Beach Bantaeng. The 2nd International Seminar on Ocean and Coastal Engineering, Environment and Natural Disaster Management, 2014, *Procedia Earth and Planetary Science,* 14(2015), 179-185.

Balley, G., King, C. P., Deves, M., Hausmann, N., Inglis, R., Laurie, E., Meredith-Williams, M., Momber, G., Winder, I., Aisharekh, A., and Sakellariou, D. (2012). Disperse: Dynamic Landscapes, Coastal Environments and Human Dispersals. http://antiquity.ac.uk/projgall/bailey334/.

BBC NEWS: http://www.bbc.com/news/world-asia-32511429 assessed on 18/07/16.

Benson, C., and Clay, E. J. (2006). Disasters, Vulnerability and the Global Economy: Implications for Less-Developed Countries and Poor Populations, in Craig S. Galbraith, Curt H. Stiles (ed.) *Developmental Entrepreneurship: Adversity, Risk, and Isolation (International Research in the Business Disciplines,* Volume 5) Emerald Group Publishing Limited, pp.115 – 145.

Davidson, B. (2016). Earthquake Break through? Leading Solar News Company challenges the world's experts in spaceweathernews.com/challenges/accessed on 18/07/2016.

Dimova. S. L. and Negro, P. (2005). Influence of construction deficiencies on the seismic response of structures. *Earthquake Engineering and Structural Dynamics*, 34 (6), 613-635.

Dreger, D. S., Gee, L., Lombard, P., Murray, M. H., and Romanowicz, B. (2005), Rapid Finite Source Analysis and Near-Fault Strong Ground Motions – Application to the 2003 Mw6.5.

Duggal S.K., *Technol. Eng.,* 448 (2007).

Facciorusso, J., Madiai, C., and Vannucchi, G. (2016). The 2012 Emilia earthquake (Italy): Geotechnical characterization and ground response analyses of the paleo-Reno river levees, *Soil Dynamics and Earthquake Engineering,* 86(2016), 71-88.

Gee, L., Neuhauser, D. Dreger, Uhrhammer, R., Pasyanos, M. and B. Romanowicz, (2003).

Gere, J.M. & Shah, H.C. (1984). *Terra Non Firma.* Freeman, New York.

Guevara L.T., Alonso J.L. Fortoul E., *Earthquake Eng.,* (1992). BS EN1998-1:2004: Eurocode 8: British Standards Institute, London.

Hermann M. Fritz., David A. Phillips., Akio Okayasu., Takenori Shimozono., Haijiang Liu., Fahad Mohammed., Vassilis Skanavis., Costas E. Synolakis and Tomoyuki Takahashi (2012). The 2011 Japan tsunami current velocity measurements from survivor videos at Kesennuma Bay using LiDAR. *Geophysical Research Letters,* VOL. 39, L00G23, doi:10.1029/ 2011GL050686, 2012.

Housner. G.W. (1970). *Strong ground motion.* Chapter 4 of Earthquake Engineering, (R.L. Wiegel, coordinating Ed), Prentice-Hall, Englewood Cliffs, NJ, 75-91.

Kramer, S.L. (1996). *Geotechnical Earthquake Engineering.* Prentice-Hall, Englewood Cliffs, NJ.

Kreimer, A., and Munasinghe, M. (1991). The Environment and Disaster Management. *Land Use Policy,* 8 (4), 269-281.

Lucero, A. (2016). Dynamic Earth: Plate Tectonics. http://platetectonicsalvin. weebly.com/ Accessed on (18/07/16).

Menshikov, V. A., Perminov, A. N., and Urlichich, Y. M. (2011). Natural Calamities and Their Forecasting: Global Aerospace Monitoring and Disaster Management. *Springer Vienna,* p 1-81.

Nelson S. A. (2014). *Natural Disasters & Assessing Hazards and Risk.* http://www.tulane.edu/~sanelson/Natural_Disasters/introduction.htm, accessed: 22-07-2016.

Olorunfemi M.O, Olarewaju V.O., and Avci M (1986). Geophysical investigation of a fault zone-Case History from Ile-Ife, southwest Nigeria. *Geophysical Prospecting,* 34(8): 1277-1284.

Olujide P.O, and Udoh A. N. (1989). Preliminary comments on the fracture systems of Nigeria. In Proceedings of the National Seminar on earthquakes in Nigeria (Ed.) Ajakaiye D. E; Ojo S. B. and Daniyan M. A. pp. 97-109.

Onosemuode C. and Dare O. T. (2010). *Environmental Research Journal,* 4(2), 187-194.

Osagie, E. O. (2008). Seismic activity in Nigeria. *Pac. J. Sci. Tech.,* 9(2): 1-6.

Press, P. and Siever, S. (2001). *Understanding Earth.* 3rd Ed. Freeman, San Francisco.

Ricketts, J.T., Loftin, M.K and Merritt, F.S. (2004). *Standard Handbook for Civil Engineers,* 5th Ed, McGraw-Hill, NY USA.

San Simeon and 2004 Mw6.0 Parkfield Earthquakes, *Seismo. Res. Lett.* 76, 40-48.

Siringoringo, D. M. and Fujino, Y. (2015). Seismic Response Analyses of an Asymmetric Base-Isolated Building During the 2011 Great East Japan (Tohoku) Earthquake. *Structural Control and Health Monitoring.* 2015; 22:71–90, Published online 11 April 2014 in Wiley Online Library (wileyonlinelibrary.com). DOI: 10.1002/stc.1661.

Sun, C., Chun, S., Ha, T., Chung, C., Kim, D. (2008). Development and application of a GIS-based tool for earthquake-induced hazard prediction. *Computers and Geotechnics,* 35(3), 436–449.

Tanikawa, H., Managi, S. and Lwin, C. M. (2014). Estimates of Lost Material Stock of Buildings and Roads Due to the Great East Japan Earthquake and Tsunami. *Journal of Industrial Ecolog.* 18(3), Pp 421-431.

The rapid earthquake data integration project, *International Handbook of Earthquake and Engineering Seismology,* Volume 81B, p 1261-1273.

Tsalha, M. S., Lar, U.A., Yakubu, T.A., Kadiri, U. A., and Duncan, D. (2015). The Review of the Historical and recent Seismic Activity in Nigeria. IOSR Journal of Applied Geology and Geophysics. 3, pp 48-56.

Uniform Building Code. Whittier, CA (1997).

USGS (2016). http://earthquake.usgs.gov/earthquakes/states/california/history.php: accessed 19/07/2016.

Vanguard Newspaper Online edition (2016). *Earth Tremor causes panic in Oyo Community.*

Vere-Jones D. (1995). Forecasting earthquakes and earthquake risks. *International Journal of forecasting,* 11 (4), 503-538.

Veronesi, F. (2015). Live Earthquake Map with Shiny and Google Map API https://1.bp.blogspot.com/S79NRcX8q6c/VWXKXHWDJhI/AAAAAAAARC0/04KKqd09DU0/s1600/Map.jpg.

Wald, D. J., Heaton, T. H. and K. W. Hudnut, K. W. (1996), The slip history of the 1994 Northridge, California, earthquake determined from strong-motion, teleseismic, GPS, and leveling data, *Bull. Seism. Soc. Am.,* 86, S49-S70.

Walker H. J. (2014). *Engineered Coasts.* Springer, Netherlands, 1-7.

Weekes (2012) *Earthquake Engineering Module,* University of Surrey, Guildford, UK.

Yeates, R.S., Sieh, k. and Allen, C.R. (1997). *The Geology of Earthquakes.* Oxford University Press, New York.

Zahi Baroudi (2016). *How to Reduce Earthquake Effects on Building.* http://structzone.com/how-to-reduce-earthquake-effects-on-buildings/.

In: Earthquakes
Editor: Wayne Coleman
ISBN: 978-1-53610-342-7

Chapter 4

PROTECTION OF CULTURAL HERITAGE BUILDINGS AGAINST EARTHQUAKES FROM A SENSITIZED STRUCTURAL ENGINEERING PERSPECTIVE

Adolfo Preciado[1,*], PhD, Osmar Rodriguez[2,†], Alejandro Ramirez-Gaytan[3,‡], PhD, Nayar Gutierrez[1,§], PhD, David Vargas[1,¶], PhD, and Gil Ochoa[1,#]

[1]Department of Housing and Urban Development, Instituto Tecnológico y de Estudios Superiores de Occidente (ITESO), Tlaquepaque, Jalisco, México

[2] Faculty of Civil Engineering, Universidad Politécnica de la Zona Metropolitana de Guadalajara (UPZMG),

[3]Department of Computer Science, University Center of Exact Sciences and Engineering, Universidad de Guadalajara (UdeG), Guadalajara, Jalisco, México

[*] adolfopreciado@iteso.mx.
[†] osmar.ubc@gmail.com.
[‡] l7m8s3r@gmail.com.
[§] nayar@iteso.mx.
[¶] davidvar@iteso.mx.
[#] gilochoa@iteso.mx.

ABSTRACT

The most common construction techniques for heritage buildings were mainly based on basic geometric rules about stability and earthquake damage observations. This situation led to the construction of massive structures with structural elements of great thickness, but still with poor seismic performance because damage was continuously presented. These restrictions have made historical buildings extremely vulnerable against earthquakes, presenting partial and total collapses. Thus, becomes evident the need of suitable technical solutions to protect them. This chapter describes the behavior and failure mechanisms of unreinforced masonry buildings and different rehabilitation and strengthening techniques. Both techniques are aimed at keeping good deformation compatibility between materials without affecting the architectonic value of the monument. Advantages and disadvantages of different rehabilitation and strengthening proposals are discussed from a sensitized structural engineering perspective taking into account that this type of buildings are very different from modern structures. These techniques aim at protecting cultural heritage against earthquakes by inducing a modified failure mechanisms that involves more energy dissipation than the original structural system.

Keywords: monuments, masonry, earthquakes, vulnerability, retrofitting, fiber reinforced polymers

1. INTRODUCTION

One of the mankind´s basic needs is housing or refuge, a fact that led ancient societies to the use of local materials such as stones, clay, and wood to protect themselves from environmental conditions and attacks by wild animals. The building techniques employed in ancient structures were later improved by means of transmitted knowledge from one generation to another that culminated with the edification of monumental castles and cathedrals. The involved construction techniques were based on geometric rules and tests on the basic stability of structural elements by trial and error, as well as the observed earthquake (EQ) damage. This led to the edification of massive structural elements of great thickness made of materials with poor behavioral properties and load capability limitations. These monumental buildings were not built to withstand EQs, but only to resist their own weight by vertical transmission to other structural elements. These limitations leave historical buildings of this kind extremely vulnerable to EQ damage.

The seismic vulnerability of these massive structures depends on three main factors: first, their great mass; second, the proximity to a potential seismic source; and third, the resistance loss over time due to material degradation. The resulting behavior may induce partial collapses or in the worst case a global failure even in static conditions without the occurrence of an EQ. Upon taking into account the extensive observed damage in cultural heritage buildings after a significant EQ, the need to find suitable technical solutions to protect them becomes evident.

The presented research is divided into two fundamental stages in an effort to describe the EQ effects on cultural heritage buildings and suggest possible solutions to protect such monuments. The first stage describes the typical behavior and failure mechanisms of unreinforced masonry structures; while the second discusses the main rehabilitation and strengthening (retrofitting) techniques in relation to (i) the compatibility of deformations among materials, and (ii) the proposal reversibility. Compatibility and reversibility are fundamental aspects in protecting cultural heritage from seismic damage and conserving as well their architectural and cultural value. Finally, the advantages and disadvantages of different rehabilitation and strengthening proposals are discussed from a sensitized structural engineering perspective, which recognizes that these massive masonry buildings are different from modern structures.

2. Seismic Vulnerability of Cultural Heritage Buildings

In the seismological context, the earth´s crust is divided into enormous tectonic plates lying on the asthenosphere, which due to pressure and temperature conditions tend to flow. The interaction mechanism between macro-plates is quite complex, similar to that of a giant jigsaw puzzle, with the occurrence of large magnitude EQs on their boundaries and very active volcanoes in some areas. Figure 1 illustrates the huge Pacific plate that spans the Pacific Ocean from Mexico and USA to New Zealand and Japan. The perimeter of the Pacific tectonic plate is known as the “ring of fire” since many of most destructive EQs in the past (seismic events of large magnitude and duration over a minute) occurred along its edge. The tectonic environment of the subduction zone of the Mexican Pacific margin (i.e., the northern part of the Middle America Trench) is a tectonically complex area of the Mexican subduction zone near to a major triple junction plate formed by the Rivera, Cocos and North American plates. The triple junction is located along the Middle America Trench offshore of Manzanillo, México. The tectonic regime may be even more complicated as the interaction between these plates has been fragmented (*i*) the overriding North American plate into two kinematically independent crustal blocks, the Jalisco and Michoacan blocks, separated by the Colima Rift (e.g., Luhr et al., 1985; Johnson and Harrison, 1990; DeMets and Stein, 1990), and (*ii*) the Cocos plate, forming a kinematically distinct plate, the northern Cocos plate, situated between the Orozco and Rivera transforms (Bandy, 1992; Stock and Lee, 1994; Bandy and Hilde, 2000 and Dougherty et al., 2012). In the Mexican region, the interaction between the Pacific plate and micro-plates such as Rivera and Cocos is complex (Ramirez-Gaytan et al., 2015). In this seismic area have occurred destructive EQs characterized by subduction mechanisms leading to multiple human loss and structural damages (e.g., 1985 Mexico City EQ Mw= 8.1 and the 2003 Colima EQ Mw= 7.5).

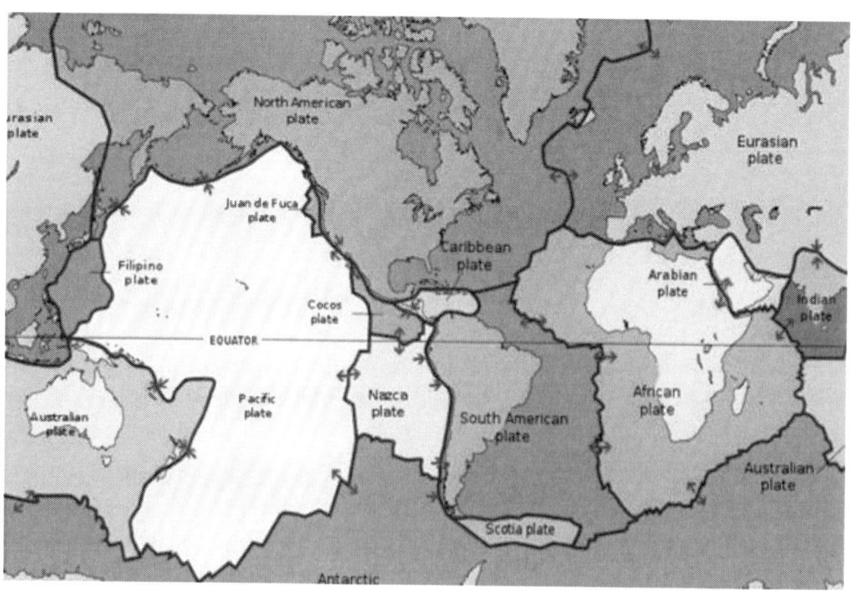

Figure 1. Tectonic plates of the Earth (USGS, 2013).

Today, technological progress in the use of global positioning systems makes it possible to monitor the generated displacements between tectonic plates to develop general estimations of stresses and probability of occurrence of catastrophic EQs. Inside the macro-plates (see Figure 1) there are micro-plates and inter-plate stressed zones (geological faults), which are characterized by the generation of EQs of medium magnitude and reduced duration. Compared to the large EQs occurred among macro-plates, the ones of micro-plates may be even more destructive due to the content of frequencies is similar to most of low rise buildings. Unfortunately, the activity of micro-plate EQs is virtually unpredictable, because most of such events occur in unknown geological faults. Examples of destructive EQs occurred on the boundaries between micro-plates include those in Sichuan, China in 2008 (Mw = 7.9 and 69200 casualties), L'Aquila, Italy in 2009 (Mw = 6.3 and 281 casualties) and more recently (2016) in central Italy, in Norcia and Amatrice (Mw = 6.2), causing the death of more than 300 people.

The risk of damage due to natural hazards like EQs is influenced by the combination between the proximity of potentially harmful seismic sources (exposure) and the intrinsic vulnerability of building´s structural components. The structural vulnerability is a characteristic property of the structure, which depends on the quality of the construction materials and the seismic design level (Preciado et al., 2015a). Analyzing the aforementioned, it may be concluded that the seismic risk is defined by the combination of three main factors (*i*) the structural vulnerability, (*ii*) the seismic hazard, and (*iii*) the building´s exposure to the seismic source. Among researchers, the seismic risk of a particular building or group of buildings is also known as seismic vulnerability.

Seismic vulnerability assessment of buildings is an issue of most importance at present time and is a concept widely used in works related to the protection of buildings. Nevertheless, there is not a rigorous and widely accepted definition of it. In general terms, vulnerability measures the amount of damage caused by an earthquake of given intensity over a structure. However, "amount of damage" and "seismic intensity" are concepts without a clear and rigorous numerical definition (Orduña et al., 2008).

2.1. Unreinforced Masonry as the Main Construction Material of Historical Buildings

Most existing cultural heritage buildings were primarily built with masonry (see Figure 2), without reinforcement or seismic design, just aimed at withstanding their self-weight vertically transmitted among structural elements. Knowledge about construction of massive masonry buildings was transmitted among constructors from generation to generation and was mainly based on trial and error. Among constructors it was well known that a building would last for many centuries if the new structural element (e.g., wall, arch, dome or vault) did not collapse when the falsework was removed. Masonry is defined as the combination of units made of natural or carved stone (Figures 2b y 3a-b) placed in different patterns with a binder (mortar). Units were usually rectangular in shape but could be of different materials, including fired-clay bricks (Figure 2a) or adobe (Figure 3d). Historically, binders consisted of clay, lime or gypsum, but after the Industrial Revolution in the beginning of the XIX century, Portland cement came into use as a high resistance component.

(a) (b)

Figure 2. Examples of mankind´s constructed cultural heritage with unreinforced masonry: (a) Roman colosseum (80 B.C.) and (b) Teotihuacan Pyramids (150 B.C.).

Compared to fabricated fired-clay masonry bricks with the use of ovens, adobe structural elements are fabricated by drying the units with the sun energy and placed together with a clay mortar. In the construction of adobe structural elements in Mexico, mainly walls (see Figure 3d), it was used as binder the same clay of units, but combined with the addition of blood, natural fibers and cattle manure to increase plasticity and provide protection against the sun and rain effects. Another type of earthen masonry structure that should be mentioned is called rammed earth. This consists in macro-blocks placed without any type of binding agent (Figure 3e). Earthen masonry walls are characterized by a reduced compressive strength if compared to the presented by stone units, therefore the roofs attached were commonly of light-weight materials (Figure 3d). Another masonry group is called dried joints and was commonly used in structural elements by the Incas in Peru (Figure 3c) and the Aztecs in Mexico. It consists of stone macro-blocks held together by gravity with no binding agent, similar to the case of rammed earth structures.

Figure 3. Different types of unreinforced masonry: (a) natural stone; (b) carved stone; (c) dried joints; (d) adobe and (e) rammed earth.

The resistance of mortars made of Portland cement is higher than that of the lime and clay mixtures used in historical buildings previously constructed to the Industrial Revolution (Meli, 1998). However, long before the use of Portland cement as a construction material, the Romans created high resistance mortars by adding volcanic ashes to lime-based binders, and so permitted with the combination of the arch the construction of monumental buildings like the Colosseum (Figure 2a), aqueducts and bridges.

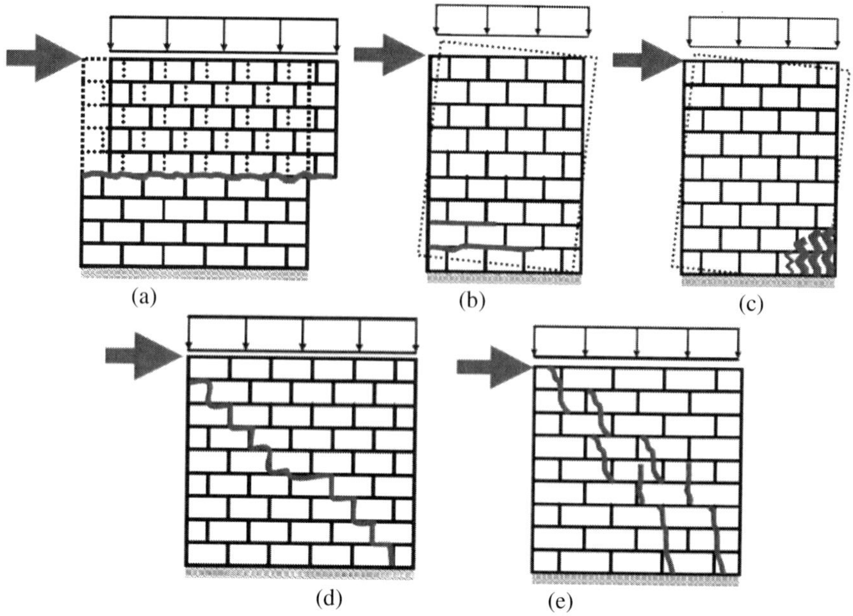

Figure 4. In-plane seismic failure modes of unreinforced masonry: (a) sliding; (b) flexion; (c) crushing; (d) stepped shear and (e) diagonal shear (Mistler, 2006).

2.2. Seismic Behavior and Typical Failure Modes of Old Masonry Buildings

In all its forms and combinations, masonry (see Figure 3) is a composite material due to the combination of units and mortar to construct the structural element. Because these two basic elements have different mechanical properties, masonry becomes a material with heterogeneous characteristics and anisotropic behavior. The anisotropy of a certain material determines its seismic response and failure mechanisms in-plane and out-of-plane, with different behavior and failure in dependency of the seismic loading direction. In the case of masonry, the units are stiffer than the mortar, but in contrast, this makes them less ductile, more brittle and limited in terms of plastic deformation. Contrary to the unit's behavior, mortar presents reduced resistance, but greater ductility with a quasi-brittle type of failure.

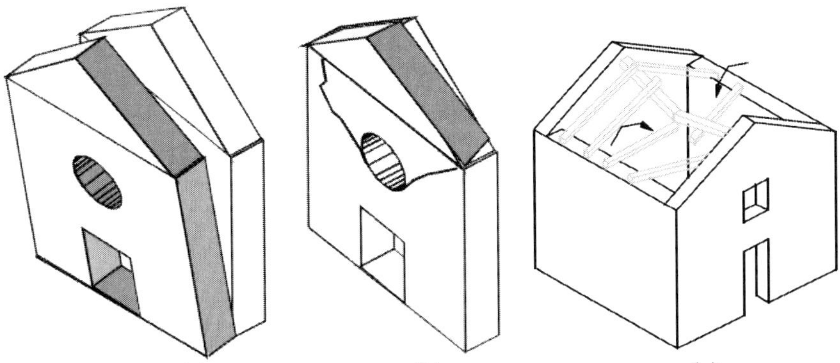

(a) (b) (c)

Figure 5. Out-of-plane seismic failure of unreinforced masonry buildings: (a) total collapse; (b) partial collapse and (c) cover collapse (D'Ayala, 2000).

Analyzing the performance of the three materials by separate, shows that as a composite, masonry presents an intermediate behavior if compared to the presented by the units and the mortar, as it is less resistant than the former but more resistant than the latter. Preciado (2011) affirms that the seismic performance and collapse mechanisms of unreinforced masonry subjected to in-plane seismic loading are determined by the material´s quality, slenderness, mass, induced shaking and vertical loading level (Figures 4-5). The anisotropic behavior of masonry is determined by its low tensile resistance and heterogeneous characteristics of its component materials. On the other hand, masonry presents a good compressive capacity in comparison to its poor tensile strength. The main collapse mechanisms that may be present in masonry walls subjected to in-plane seismic loading are due to shear (diagonal tension) and flexion stresses as illustrated in Figure 4. Torsion may occur in case of slender structures and buildings with strong irregularities in plan. Normally, the roofing of historical buildings are classified as "light" or "heavy" systems, depending on the construction material and used structural elements. Light covers are made of timber beams and clay tiles, while heavy systems include such masonry elements as domes, vaults and arches. These two systems behave quite distinctly when subjected to EQs. In case of timber systems, the beams hit the walls in different patterns that may induce the failure of the resistant element by overturning (see Figure 5a-b).

Neither light nor heavy roofing systems behave as a rigid diaphragm as in the case of most recent structures constructed with reinforced concrete slabs or steel-deck systems. Lateral seismic inertia forces presented out of the plane of the structural element may induce a different failure mode if compared to the presented in-plane due to the anisotropic behavior of masonry. Thus, the structural element may present a complete (Figure 5a) or partial (Figure 5b) collapse by overturning due to the limited resistant moment and slenderness of the walls.

(a) (b)

Figure 6. Seismic damage in cultural heritage buildings: (a) 2011 M6.3 Christchurch EQ and (b) 2009 M6.3 L'Aquila EQ (Preciado et al., 2015a).

When an EQ strikes, the massive roofing systems of cathedrals and churches tend to open at their support points due to the tremendous cover´s weight. Those openings may induce lateral out-of-plane inertia forces, leading to partial collapses or a complete failure of the entire roofing system (see Figs. 5c and 6b). It is worth noting in Figure 6 the EQ damages in historical buildings by the transmission of elevated inertia forces in different planes. More specifically, in Figure 6a may be observed a complete failure of the main façade and timber roofing system, while Figure 6b illustrates a total collapse of the masonry cover into the interior part of the church along with the collapse of almost the entire building by combination of stresses. These damages were generated by a Mw = 6.3 EQ in two different parts of the world, one in New Zealand (Figure 6a) and the other one in Italy (Figure 6b).

2.3. Sudden Collapse of Massive Old Buildings in Static Conditions

Another unusual situation that may induce a total collapse in historical unreinforced masonry buildings is the degradation of the material´s inherent resistance combined with differential soil settlements. This complete failure may occur even in static conditions without any induced vibration. The heavy mass of these monumental buildings tends to concentrate most of the compressive stresses at the base. This concentration, combined with the material´s degradation due to weathering and humidity over time, may trigger a complete collapse by exceeding the structure´s intrinsic compressive strength. The fragility and potential explosiveness of this mechanism causes masonry defragmentation, being extremely dangerous for people inside the building or in the surrounding area (see Figure 7b). Such sudden collapses have occurred for centuries, as in the case of the Bell Tower of *Piazza San Marco* in Venice, Italy in 1902. That failure was attributed to a combination of differential settlements, strength degradation and the great mass of the tower. A replica was built to replace the original one as shown in Figure 7a. A more recent case is the presented by the Civic Tower of Pavia, Italy in 1989 as depicted in Figure 7b; as well as the presented by the bell tower of the St. Mary Magdalene Church in Goch, Germany in 1992.

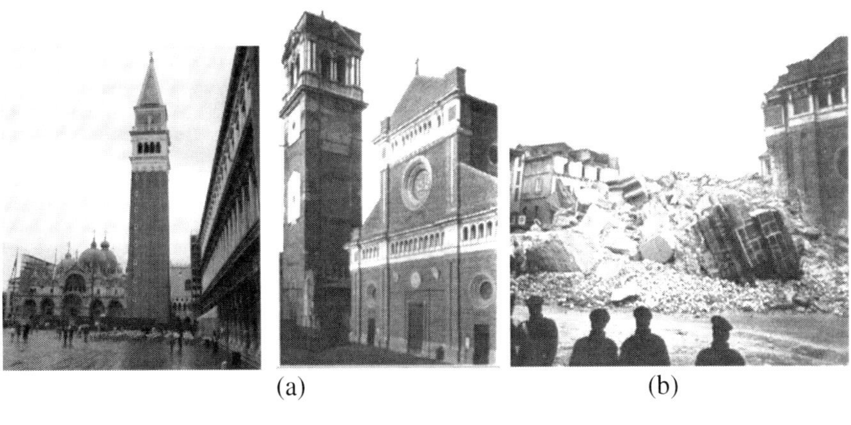

(a) (b)

Figure 7. Examples of sudden collapses of monumental buildings: (a) replica of the Bell Tower of *Piazza San Marco* and (b) the Civic Tower of Pavia (Binda, 2008).

The sudden collapses of both historical masonry monuments caught the attention of many researchers and authorities around the world who began to seek possible explanations for those unfortunate events. For more information on this topic, the interested reader should consult the developed research by Binda et al., (1992), Macchi (1993), GES (1993) and Binda (2008).

3. Seismic Protection of Cultural Heritage

Seismic protection of existing cultural heritage buildings is a topic of intensive research and great interest for many researchers. At present, there is no specific intervention technique able to prevent the partial or total collapse of such buildings, much less to predict or prevent the occurrence of EQs. Each historical building is unique and represents a challenge for anyone who pretends to protect it against natural and human-generated risks. This may be even more complex due to the EQ characteristics in terms of (*i*) wave propagation, (*ii*) seismic source mechanisms, (*iii*) content of frequencies, (*iv*) duration, and (*v*) soil site effects. Most of structural engineers use to rehabilitate or retrofit existing modern reinforced concrete or steel buildings and so attempt to apply the same techniques to the brittle masonry of historical constructions. Their intervention solutions commonly involve high-resistant conventional materials that are not compatible with ancient masonry.

The high differences between material´s deformations induce in EQ conditions a concentration of stresses that may cause a partial failure mechanism of vulnerable structural parts, or in the worst case a complete collapse of the monument. Moreover, the involved structural engineer usually does not know the history and architectural value of the monument under intervention, and so tends to use reinforced concrete insertions (confinement) and adjoining walls. These irreversible inserted elements and the use of plaster are no longer recommended for the rehabilitation or retrofitting of cultural heritage buildings with bare-stone works. Therefore, what is urgently needed is a sensitized structural engineering perspective for the seismic protection of our existing cultural heritage constructions.

As a first stage, it is recommended the development of a historical analysis of the used construction materials and presented damages in the building and restorations. Furthermore, to carry out non-destructive experimental tests *in situ* and in the laboratory to determine the mechanical properties of the component materials. The dynamic properties of the building such as fundamental frequency and vibration modes can be obtained by experimental *in situ* tests with accelerometers located in different parts of the structure, and by modal dynamic analyses using the numerical 3D model. This mechanical and dynamical information is fundamental in determining the seismic strengthening proposal of the building under study, but must be accompanied by the historical analysis, a thorough characterization of the nearest seismic source(s), and a detailed structural analysis.

3.1. Rehabilitation of Cultural Heritage Buildings

Rehabilitation (restoration) works combine materials with similar characteristics and appearance to those of the original ones and the application of typical construction techniques to correct localized damage in order to keep the building in good conditions. The most common activities in rehabilitation works involve replacing damaged units and mortar (Figure 8a), as well as the sealing of cracks (Figure 8b) with mortar of Portland cement. Sealing the structural elements prevents cracks from propagating and water from infiltrating the masonry due to weathering. Replacing units is not an easy work, this is due to the fact that in most cases the origin of the damaged units is unknown and it is difficult to match the color of new units to the original ones. One solution may be the treatment of the new pieces with induced humidity and weathering aimed at reducing the resistance to produce aged units similar to the originals.

(a) (b)

Figure 8. Rehabilitation of apparent masonry walls: (a) mortar repointing and replacement of damaged units and (b) sealing of cracks with polymers.

Simulating aspects like color and aging may not be too difficult; the main problem is to match the mechanical properties of the new and original units, but in most cases this is not achieved or even taken into account. However, both rehabilitation techniques, substitution of units and mortar (see Figure 8) may ensure a good compatibility among deformations in order to prevent the concentration of stresses under EQ. Bednarz and Bajno (2014) propose using mortars of polymers combined with natural fibers because this may be ecologically effective and provide good deformation compatibility because of its reduced elasticity modulus compared to that of conventional materials such as Portland cement-based mortars. The sealing of cracks technique with polymers has been used in several historical buildings around the world, as in the case of the Kaisariani monument in Athens, Greece. Moropoulou et al., (2016) developed a detailed research on this monument by means of numerical investigations to optimize the use of mortars with good resistance that were compatible with the original ones.

In case of historical buildings located inside large metropolises that are exposed to pollution, we find that this may generate an aging state of the bare-carved-stone similar to the induced by weathering, as illustrated in Figure 9. Years of high humidity and smog, as well as the excrement from bats and pigeons are the fundamental factors that contribute to masonry aging. In recent years, material degradation caused by atmospheric conditions and the rehabilitation of cultural heritage buildings have become topics of intensive research (see Lourenco et al., 2014; Ravelo 2011; Ordaz 2004 and Pérez 2000).

Previously, abrasive chemical products were commonly used to remove stains and marks of humidity from the bare-stone of masonry buildings. But today, biodegradable products have caught the attention of conservationists and researchers. Figure 9a shows the rehabilitation of a badly-stained historical masonry building in New York using biological products, as well as the application to extremely-damaged stoneworks in cemetery crypts (Figure 9b). It is worth noting in the rehabilitation works shown in Figure 9 that the original color of the bare-stone may be successfully obtained with these new biological solutions.

In general terms, the main objective of rehabilitating cultural heritage buildings is to preserve them from invasive vegetation, humidity and aging, as well as the sealing of cracks and renovation of damaged units (Figures 8 and 9). As aforementioned, cultural heritage buildings were not designed to resist laterally-transmitted inertia forces caused by EQs, but were constructed only to withstand their own vertical weight. Rehabilitation works pretend to recover the vertical load-bearing capacity of structural elements and keep the monument in its original conditions and good conservation state, without increasing the lateral load capacity against EQs.

(a) (b)

Figure 9. Rehabilitation of stained stone masonry with biological products: (a) New York Public Library and (b) application in old cemetery crypts (LIW, 2016).

3.2. Seismic Strengthening of Cultural Heritage Buildings

Compared to the rehabilitation of a cultural heritage building, strengthening (or retrofitting) procedure aims at using advanced structural engineering techniques combined with high-performance materials in an effort to substantially improve seismic performance. High-performance materials (smart materials) have especial characteristics and behavior different from those presented by the conventional ones. Thus, their failure mechanisms are distinct, though they have similar or greater strength with reduced elasticity modulus. This balance between stiffness reduction and good resistance may help to improve the compatibility of deformations between the original main material and the retrofitting proposal. These achievements may create an energy-dissipative system able to withstand and release lateral inertia forces through a ductile behavior and improved failure mechanism. Thus, the structure may fail ductile, instead of collapsing brittle or quasi-brittle. The level of energy dissipation is assessed by the enclosed area in the envelope of the capacity curve (lateral force *vs.* displacement) shown in Figure 12b.

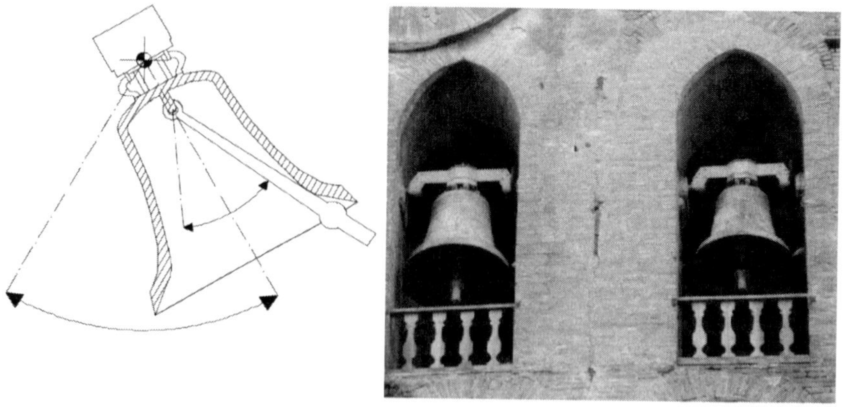

Figure 10. Vertical cracking in a tower induced by bells swinging (Beconcini et al., 2001).

As stated in Section 2, historical masonry buildings are extremely vulnerable to EQ damage because they were conceived to only resist their own vertical weight. Therefore, implementing structural strengthening must be based not only on detailed studies like those developed for rehabilitation works, but also on extensive evaluations of various retrofitting proposals to determine which is the most suitable in a specific case. Any strengthening proposal has to be compatible with the original structure in terms of deformations, and must be both durable and reversible; i.e., it can be removed if it shows deficiencies, or if a better solution arises. A second recommendation is to apply the seismic retrofitting simultaneously with rehabilitation works. The first stage of this process would involve identifying the zones with concentrations of masses in different areas of the building that may increase the EQ inertia forces in order to prevent extensive damage. A mass-reduction strategy can be implemented in floor systems by removing earthen infills and replacing timber beams with light clay tiles or hardwood flooring. It is also advisable to move large, concentrated masses like pianos, furniture and various non-structural elements from the chorus or other elevated levels to the ground floor.

Most masonry towers of churches and cathedrals shelter at belfry (space for the ringing bells) enormous bells made of heavy iron that often exceed 1m in diameter. When the bells ring, there is a transmission of elevated vertical and horizontal forces to the supporting walls that may cause strong damage to the masonry, as shown in Figure 10. This initial cracking may later be the trigger of a complete failure mechanisms under EQ conditions. For this reason, in some places of Europe is being implemented isolated bell-ringing systems with steel frames in order to transmit the induced forces to the foundation without affecting the masonry. In addition to protecting the masonry, this system has also the advantage of synchronizing the ringing of bells with an external clock.

(a) (b) (c)

Figure 11. Existing applications of old prestressing: (a) external view of a bell tower; (b) internal view of the tower and (c) prestressed arches and vaults.

Achieving good deformation compatibility among materials during seismic retrofitting interventions of cultural heritage buildings is essential in preventing a concentration of stresses that may strongly damage the structure. Moreover, as mentioned above, reversibility is a key aspect in order to avoid permanent insertions and damages to the architectural beauty of the building. The durability of some conventional materials used to strengthen buildings is unknown and difficult to determine, this is due to the lack of existing evidence if compared to masonry, earthen and timber structures. Conventional materials have less durability than those just mentioned because they are vulnerable to weathering, presenting corrosion due to excess of humidity and salinity, and may be subject to brittle failure in case of fire. Steel structures are more vulnerable to fire than reinforced concrete ones due to the material intrinsic characteristics.

Figure 11 shows some existing applications of old prestressed iron bars in old masonry towers and vaults in several Italian cities, including Rome, San Gimignano and Pisa. The external part of the tower (Figure 11a) shows the diagonal anchorages of the old prestressing system. In the case of vaults, the anchorage plates are horizontally located (Figure 11b), while in the case of long bars, a tightener that connects two iron bars is visible in the central part. Heating those iron bars generated dilatation and by cooling the prestressing effect was applied at the anchorages.

Preciado et al., (2015b) mention that the prestressing technique was first applied in the beginning of the XIX century as illustrated in the masonry structures of Figure 11. The modern prestressing technique that uses high-resistance steel and concrete is based on the old technique and its use dates to the late XX century. This technique has multiple uses, from long-span bridges to high resistance vertical structures. Existing old prestressing is still commonly visible in masonry buildings in Europe. Its purpose is to prevent the brittle overturning of walls when exposed to lateral seismic forces. This rudimentary strengthening technique presented acceptable performance in preventing partial collapses, but in-plane damage was visible after every considerable EQ because there was no effective increase of the building´s capacity.

The prestressing technique is an effective retrofitting measure if it is vertically applied (Sperbeck, 2009 and Preciado, 2011), this ensures a seismic energy dissipation by increasing lateral force and displacement capacity by modifying the failure mechanism (see Figure 12). It is advisable to apply the prestressing device (tendon and anchorage plates) after calculating the pre-compression forces to be applied in specific parts of the structure, taking into account detailed nonlinear seismic analyses (Preciado et al., 2016a-b).

New materials, such as fiber reinforced polymers (FRPs), that behave differently from conventional ones, have shown to be usable in the seismic retrofitting of masonry and reinforced concrete structures (Ascione et al., 2005; D'Ambrisi et al., 2013a and b; Foraboschi and Vanin, 2013b; Muciaccia and Biolzi, 2012 and Fedele et al., 2014). FRPs are fabricated with glass, carbon and aramid fibers in the shape of tendons, laminates and meshes. They offer corrosion protection and high resistances with reduced elasticity modulus. It is also possible to include optical fiber sensors inside the FRP devices to allow with the instrumentation the monitoring of buildings. These smart materials are much lighter than conventional steel tendons, but in contrast, are vulnerable to brittle failure under fire conditions or ultimate stress.

(a)

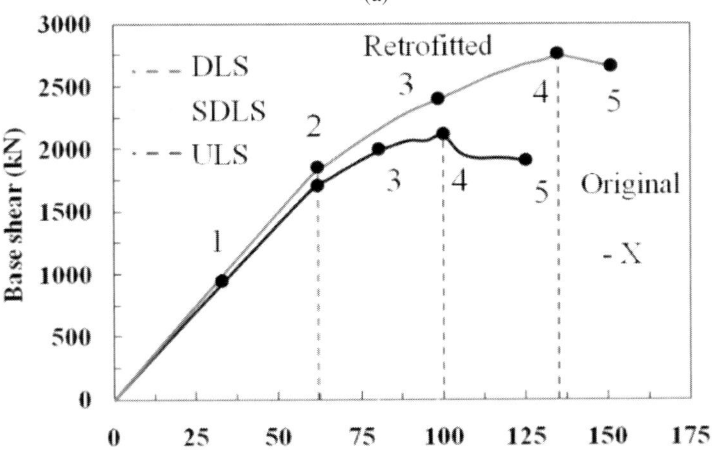

(b)

Figure 12. Reversible FRP prestressing of a historical tower: (a) Cathedral of Colima, Mexico and (b) seismic performance comparative (Preciado et al., 2016a).

Another smart material made of metallic alloys such as nickel and titanium (Nitinol) is also known among researchers as shape memory alloy (SMA), because it presents an inverse behavior to that of conventional materials. This material can undergo extensive deformation without presenting plastic damage, and its loading/unloading envelope dissipates large amounts of energy. This energy-dissipation property similar to the action of a damper has found interesting applications in the seismic protection of cultural heritage buildings (e.g., Indirli et al., 2001; Castellano, 2001; Desroches and Smith, 2003; Fugazza, 2003 and Preciado, 2011). Figure 13a illustrates applications of the FRP-wrapping of walls with laminates, as well as different FRP tendons used in prestressing devices (Figure 13b). Both retrofitting techniques with FRP tendons and laminates may effectively increase the seismic performance of masonry structures. The shear and flexion enhancement they provide increases the masonry´s lateral load and displacement capacity, thus converting the structure into a high-energy dissipative system.

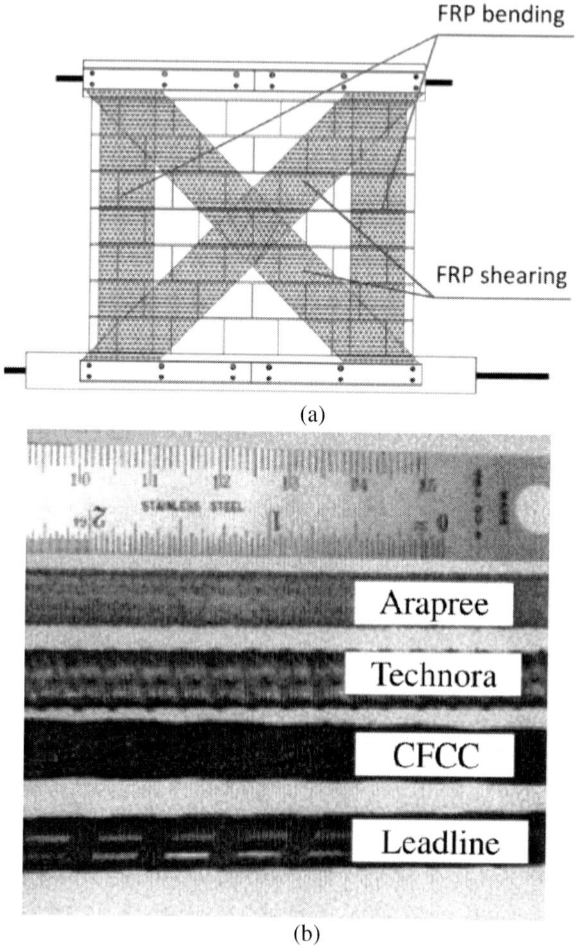

Figure 13. Seismic strengthening of masonry structures: (a) Carbon FRP laminates (Bischof y Suter, 2014) and (b) FFP tendons (Zhang et al., 2001).

Conventional confinement insertions may be useful in cases of historical building without a bare-surface, as is the case of most historical buildings in Mexico, which have plaster and paint on their façades and in internal areas of churches and cathedrals as a result of EQ damage repairs. Strengthening of structures by confinements is widely used in Latin America because it is a fast and economical solution. FRP meshes may also be effective in retrofitting walls, domes and vaults, but require the use of plaster, which converts this strengthening solution into a non-reversible proposal.

4. Future Perspective

The use of steel elements (rebar and prestressing tendons) is not recommended as a strengthening measure for ancient brittle masonry structures due to the great difference in deformation compatibility and vulnerability to corrosion. Recent trends in the seismic retrofitting of historical structures by prestressing involve the use of smart materials with a low elasticity modulus and high resistances such as FRP and Nitinol SMAs. These futuristic materials are continuously being investigated by adding new alloys and fibers. The main objective is to create new materials that differently behave from conventional ones. Sensors can be installed inside FRP devices for monitoring the induced vibrations and the possibility of a future calibration of the prestressing forces. Monitoring control may also allow the set-up of EQ early-warning systems and actuators capable of activating prestressing forces, but only in case of reaching a certain level of shaking. The prestressing system may be passive most of the time and then become activated by means of the actuator, only during EQ conditions. The suitability and effectiveness of early-warning systems strongly depend on the EQ source, directivity and source location. For example, Mexico City´s early-warning system allows a time of about one minute for EQs that occur on the western Pacific coast.

Conclusion

Unreinforced masonry is the main construction material of cultural heritage buildings. While it presents an acceptable compressive resistance, it only allowed the construction of massive monumental buildings aimed at resisting only their own weight through a system of vertical resistance based on walls and columns. In contrast, the tensile strength of masonry is very low, ranging between 5-10% of its compressive strength. This limitation makes masonry buildings extremely vulnerable to the presence of shear, torsion and flexion stresses induced by EQ shaking. Ringing of bells may also induce horizontal shaking and micro-cracking to supporting walls and may trigger partial or complete failure mechanisms in case of strong EQs.

The structural vulnerability of cultural heritage buildings is increased by the tremendous mass of these monuments and the enormous openings that characterize them. In case of EQ, the level of inertia forces is proportional to the mass, inducing strong shaking to the different parts of the building. The large openings concentrate most of the stresses, thus, may behave as failure points leading to a complete failure. Seismic vulnerability is also increased by the proximity of the building to a potential seismic source that may generate considerable on-site intensities. Moreover, the poor conservation state, weathering, induced vibration by traffic, and the lack of interest of authorities, all combine to place cultural heritage buildings at high risk.

For protecting historical buildings, our recommendations include conducting detailed studies of the characteristics of potential seismic sources, historical analyses, evaluations of damage and restorations and non-destructive experimental tests to determine the structure´s mechanical and dynamical properties. This information is the basis to propose the most suitable rehabilitation and seismic-strengthening measures, always taking into account a sensitized structural engineering perspective. These improvement measures are aimed at achieving a better seismic performance by generating a highly-efficient EQ energy-dissipative mechanism. Smart materials have proven to be very useful for the seismic retrofitting of masonry and reinforced concrete buildings, but a huge field of study remains to be explored. These futuristic materials behave differently from conventional ones, and may be capable of dissipating large amounts of EQ energy with good deformation compatibility among materials. Early-warning systems constitute a topic of great interest to researchers in terms of future applications to prevent human casualties in case of strong EQs. Finally, sensors can be installed inside FRP tendons and meshes for vibration monitoring, future calibration and activation of prestressing forces combined with EQ warning systems.

REFERENCES

Ascione, L., Feo, L. and Fraternali, F (2005). "Load carrying capacity of 2D FRP/strengthened masonry structures." *Composites Part B: Engineering*, 36(8): 619–26.

Bandy, W. L. (1992). "Geological and geophysical investigation of the Rivera-Cocos plate boundary: implications for plate fragmentation." PhD Dissertation, Texas A&M University, College Station, Texas.

Bandy, W. L. and Hilde, T. W. C. (2000). "Morphology and recent history of the ridge propagator system located at 18ºN, 106ºW." *The Geological Society of America* 334: 29-40.

Beconcini, M. L., Bennati, S. and Salvatore, W. (2001). "Structural characterization of a medieval bell tower: First historical, experimental and numerical investigations." Proceedings of the 3rd International Congress on Structural Analysis of Historical Constructions (SAHC), Guimaraes, Portugal.

Bednarz, L., and Bajno, D. (2014). "Remains of urban heritage defence structures conservation, monitoring and use." *International Conference on Structural Analysis of Historical Constructions* (SAHC). Mexico City.

Binda, L., Gatti, G., Mangano, G., Poggi, C. and Sacchi-Landriani, G. (1992). "The collapse of the civic tower of Pavia: A survey of the materials and structure." *Masonry International* 11-20.

Binda, L. (2008). "Learning from failure: Long-term behaviour of heavy masonry structures." Polytechnic of Milano, Italy. Published by WIT press, GB.

Bischof, P. and Suter, R. (2014). "Retrofitting Masonry Walls with Carbon Mesh." *Polymers*, 6(2): 280-299.

Castellano, M. G. (2001). "Innovative technologies for earthquake protection of architectural heritage." Proceedings of the International Millennium Congress: More than two thousand years in the history of Architecture, UNESCO-ICOMOS, Paris, France.

D'Ambrisi, A., Focacci, F. and Caporale, A. (2013a). "Strengthening of masonry-unreinforced concrete railway bridges with PBO-FRCM materials." *Composite Structures*, 102: 193–204.

D´Ambrisi, A., Feo, L. and Focacci, F. (2013b). "Experimental and analytical investigation on bond between Carbon-FRCM materials and masonry." *Composite Structures*, 46: 15–20.

D'Ayala, D. (2000). "Establishing correlation between vulnerability and damage survey for churches." Proceedings of the 12th World Conference on Earthquake Engineering (12WCEE), 30 January - 4 February, 2000, Auckland, New Zealand.

DeMets, C. and Stein, S. (1990). "Present-day kinematics of the Rivera plate and implications for tectonics of southwestern Mexico." *Journal of Geophysical Research* 95:21931-21948.

Desroches, R. and Smith, B. (2003). "Shape memory alloys in seismic resistant design and retrofit: A critical review of the state of the art, potential and limitations." School of Civil and Environmental Engineering, *Georgia Institute of Technology*, Atlanta, USA.

Dougherty, S. L., Clayton, R. W. and Helmberger, D. V. (2012). "Seismic structure in central México: implications for fragmentation of the subducted Cocos plate." *Journal of Geophysical Research* 117:B09316, doi: 10.1029/2012JB009528.

Fedele, R., Scaioni, M., Barazzetti, L., Rosati, G. and Biolzi, L. (2014). "Delamination tests on CFRP-reinforced masonry pillars: Optical monitoring and mechanical modeling." *Cement and Concrete Composites*, 45: 243-254.

Foraboschi, P. and Vanin, A. (2013b). "New methods for bonding FRP strips onto masonry structures: experimental results and analytical evaluations." Composites: Mechanics, Computations, Applications, *An International Journal*, 4(1): 1-23.

Fugazza, D. (2003). "Shape-memory alloy devices in earthquake engineering: Mechanical properties, constitutive modeling and numerical simulations." Master thesis, University of Pavia, Italy.

GES (1993). "Technical opinion about the collapse of the bell tower of St. Maria Magdalena in Goch, Germany." Gantert Engineering Studio.

Indirli, M., Castellano, M. G., Clemente, P. and Martelli, A. (2001). "Demo-application of Shape Memory Alloy Devices: The rehabilitation of the S. Giorgio church bell-tower." *Proceedings of SPIE, Smart Structures and Materials*.

Johnson, C. A. and Harrison, C. G. A. (1990). "Neotectonics in central México." *Physics of the Earth and Planetary Interiors*, 64: 187–210.

LIW (2016). "Applications of de D/2 biological solution on historical masonry structures and buildings in USA." Lime Works Inc., USA.

Lourenco, P. B., Van Hees, R., Fernandes, F., Lubelli, B. (2014). "Characterization and damage of brick masonry." Book Part, Building *Pathology and Rehabilitation*, vol. 2.

Luhr, J. F., Nelson, S. A., Allan, J. F. and Carmichel, I. S. (1985). "Active rifting in southwestern Mexico: manifestations of an incipient eastward spreading-ridge jump." *Geology* 13: 54-57.

Macchi, G. (1993). "Monitoring medieval structures in Pavia." *Structural Engineering International*, I/93.

Meli, R. (1998). "Structural engineering of the historical buildings (in Spanish)." México: Civil Engineers Association (ICA) Foundation, A. C.

Moropoulou, A., Apostolopoulou, M., Moundoulas, P., Aggelakopoulou, E., Siouta, L., Bakolas, A., Douvika, M., Karakitsios, P., Panagiotis G. Asteris. (2016). "The Role of Restoration Mortars in the Earthquake Protection of the Kaisariani Monastery." ECCOMAS Congress, Crete Island, Greece.

Mistler, M. (2006). "Deformation-based seismic measure concept for masonry constructions (in German)." Doctoral thesis, University of Aachen, Germany.

Muciaccia, G. and Biolzi, L. (2012). "Thermal degradation of fiber reinforced extruded materials." *Fire Safety Journal*, 49(4): 89–99.

Ordaz, M. (2004). "Arquitectura Religiosa Virreinal de Yucatán. El conocimiento histórico-técnico de las iglesias con estructura espacial conventual." Tesis Doctoral, Universidad Politécnica de Cataluña.

Pérez, L. (2000) "Influencia del medio ambiente en la patología de los monumentos de alto valor histórico construidos con materiales pétreos naturales. Centro histórico de la habana." Tesis Doctoral, Habana, Cuba.

Orduña, A., Preciado, A., Galván, J. F. and Araiza, J. C. (2008), "Vulnerability assessment of churches at Colima by 3D limit analysis models." *Proceedings of the 6th International Conference on Structural Analysis of Historical Constructions* (SAHC), Bath, UK.

Preciado A. (2011). "Seismic vulnerability reduction of historical masonry towers by external prestressing devices." Doctoral thesis, Technical University of Braunschweig, Germany and University of Florence, Italy.

Preciado, A., Ramirez-Gaytan, A., Salido-Ruiz, R. A., Caro-Becerra, J. L. and Lujan-Godinez, R. (2015a). "Earthquake Risk Assessment Methods of Unreinforced Masonry Structures: Hazard and Vulnerability." *Earthquakes and Structures*, 9(4): 719-733.

Preciado, A., Bartoli, G. and Budelmann, H. (2015b). "The Use of Prestressing Through Time as Seismic Retrofitting of Historical Masonry Constructions: Past, Present and Future Perspective." *Ciencia Ergo-Sum*, 22(3): 242-252.

Preciado, A., Budelmann, H. and Bartoli, G. (2016a). "Earthquake protection of colonial bell towers in Colima, Mexico with externally prestressed FRPs." *International Journal of Architectural Heritage*, 10(4): 499-515.

Preciado, A., Sperbeck, S. T. and Ramirez-Gaytan, A. (2016b). "Seismic vulnerability enhancement of medieval and masonry bell towers externally prestressed with unbonded smart tendons." *Engineering Structures*, 122: 50-61.

Ramirez-Gaytan, A., Bandy, W. L., Jaimes, M. A., Salido-Ruiz, R. A., Preciado, A., Huérfano, V. and Cárdenas-Monroy, C. (2015). "Analysis of the unusual earthquake of 13 August 2006 in Michoacán, México." *Volcanology and Seismology*, 9(6): 412-428.

Ravelo, G. (2011). "Influencia de los elementos climáticos en el deterioro de las fachadas de edificaciones del barrio Colón." Arquitectura y Urbanismo, vol. XXXII, No. 3.

Sperbeck, S. (2009). "Seismic risk assessment of masonry walls and risk reduction by means of prestressing." Doctoral thesis, Technical University of Braunschweig, Germany and University of Florence, Italy.

Stock, J. M. and Lee, J. (1994). "Do microplates in subduction zones leave a geological record?." Tectonics 13:1472-1487.

USGS (2013). "Understanding plate motions." *United States Geological Survey.*

Zhang, B., Benmokrane, B., Chennouf, A., Mukhopadhyaya, P. and El-Safty, A. (2001). "Tensile behavior of FRP tendons for prestressed ground anchors." *Composites for Construction*, 5(2): 85-93.

BIOGRAPHICAL SKETCH

Adolfo Preciado

Affiliation: Western Institute of Technology and Higher Education (ITESO)

Education: BEng, MEng, PhD

Business Address: Periferico Sur Manuel Gómez Morin 8585, 45604 Tlaquepaque, Jalisco, México

Research and Professional Experience:

Civil Engineer graduated with honors by the University of Colima (2004) and Master in Civil Engineering (Seismic Protection of Buildings) in 2007 by the same University. In 2011 obtained de double PhD degree in Civil Engineering by the University of Florence, Italy and the Technical University of Braunschweig, Germany in Risk Management of Natural and Man Induced Hazards on Structures and Infrastructure. In 2012 developed a two years Postdoctoral research stay at the University of Auckland, New Zealand about the seismic strengthening of historical buildings damaged by the 2010 and 2011 Canterbury earthquakes. From 2014 to the beginning of 2016 worked as a Professor-Researcher in the Polytechnical University of Guadalajara and recently is a Professor and Researcher in the DHDU of ITESO. Dr. Preciado has more than 10 years of research experience and belongs to the National System of Researchers (SNI).

Professional Appointments:

- Postdoc and invited Professor at the Faculty of Civil Engineering of the University of Auckland (2012-2013)
- Professor and Researcher at the Faculty of Civil Engineering of the Polytechnical University of Guadalajara (UPZMG) (2014-2016)
- Professor and Researcher at the Faculty of Civil Engineering of the Department of Habitat and Urban Development, Western Institute of Technology and Higher Education (ITESO) (2016-Act.)

Honors:

- Civil Engineer graduated with honors by the University of Colima (2004)
- Member of the National System of Researchers (SNI), CONACYT, Mexico since 2016.

Publications:

Selected Publications:

Preciado, A., Sperbeck, S. T. and Ramirez-Gaytan, A. (2016). "Seismic vulnerability enhancement of medieval and masonry bell towers externally prestressed with unbonded smart tendons". *Engineering Structures*, 122: pp. 50-61.

Preciado, A., Budelmann, H. and Bartoli, G. (2016). "Earthquake protection of colonial bell towers in Colima, Mexico with externally prestressed FRPs". *International Journal of Architectural Heritage*, 10(4), pp. 499-515.

Preciado, A., Orduña, A, Bartoli, G. and Budelmann, H. (2015). "Façade seismic failure simulation of an old Cathedral in Colima, Mexico by 3D Limit Analysis and Nonlinear Finite Element Method". *Engineering Failure Analysis,* 49: pp. 20-30.

Preciado, A., Bartoli, G. and Budelmann, H. (2015). "Fundamental aspects on the seismic vulnerability of ancient masonry towers and retrofitting techniques". *Earthquakes and Structures*, 9(2): pp. 339-352.

Ramirez-Gaytan, A., Bandy, W. L., Jaimes, M. A., Salido-Ruiz, R. A., Preciado, A., Huerfano, V. and Cárdenas-Monroy, C. (2015). "Analysis of the unusual earthquake of 13 August 2006 in Michoacán, México". *Volcanology and Seismology*, 9(6): pp. 412-428.

Preciado, A., Bartoli, G. and Budelmann, H. (2015). "The Use of Prestressing Through Time as Seismic Retrofitting of Historical Masonry Constructions: Past, Present and Future Perspective". *Ciencia Ergo-Sum*, 22(3): pp. 242-252.

Preciado, A. (2015). "Seismic vulnerability and failure modes simulation of ancient masonry towers by validated virtual finite element models." *Engineering Failure Analysis*, 57: pp. 72-87.

Preciado, A., Ramirez-Gaytan, A., Salido-Ruiz, R. A., Caro-Becerra, J. L. and Lujan-Godinez, R. (2015). "Earthquake Risk Assessment Methods of Unreinforced Masonry Structures: Hazard and Vulnerability". *Earthquakes and Structures*, 9(4): pp. 719-733.

In: Earthquakes
Editor: Wayne Coleman

ISBN: 978-1-53610-342-7

Chapter 5

SEISMIC RISK KNOWLEDGE SOURCES AND PERCEPTION: THE CASE OF PREPARATORY SCHOOL STUDENTS IN MEXICO CITY

Jaime Santos-Reyes**, *Tatiana Gouzeva* and *Galdino Santos-Reyes
Grupo de Investigación SARACS, ESIME, ZAC.,
Instituto Politecnico Nacional, México

ABSTRACT

The chapter presents the results related to the preferred sources of information related to Risk, Hazard and Natural disaster. Also, some results of seismic risk perception and the preferred sources of education on seismic risk are presented. The sample size of the cross-sectional study was N = 1489 that included six Preparatory school located in Mexico City. The main findings of the study have been the following: a) overall, the results show that the students are well aware of the words 'Hazard' (99.6%, Mean = 2.00, SD = 0.063), 'Risk' (99.5%, Mean = 2.00, SD = 0.069), and 'Natural disaster' (99.2%, Mean = 2.00, SD = 0.118); b) the main source of information associated with the term 'Risk' (Mean = 3.27,

* Corresponding author: Tel.: +52 55 57296000, Ext. 54663; Fax: +52 55 57296000, Ext. 54588. Email address: jsantosr@ipn.mx; jrsantosr@hotmail.com (J. Santos-Reyes).

SD = 0.891) was the "School." The term 'Natural disaster', on the other hand, has been found to be "TV" (Mean = 1.87, SD = 0.334); c) regarding the participants' knowledge on the three seismic zones of the Capital City, the overwhelming majority of them did not know this fact (89.1%, Mean = 1.10, SD = 0.294); d) students perceived that an earthquake could occur "At any time" (86.9%, Mean = 4.72, SD = 0.816); e) students prefer to learn more on seismic risk by the following sources and in this particular order: "Civil protection" (44%), "School" (22%), and "Internet" (20%); f) In general, the participants of the study showed a lack of knowledge on seismic risk (see c) above), even though, they are well aware that an earthquake could happen at any time (see d) above). It is hoped that the preliminary results presented here may serve as the basis for a more in-depth analysis, for example, of mass media, social networks related to earthquakes, in the context of the Mexican culture. The paper gives an account of an ongoing research project.

Keywords: preparatory school, earthquakes, knowledge and education, seismic risk

INTRODUCTION

In recent years, disasters triggered by earthquakes have highlighted the vulnerability of communities living in seismic regions on Earth (Beck, André-Poyaud, Paule-Annick, Chardonnel & Lutoff, 2012; Blaikie, Cannon, Davis & Wisner, 1994; Sims, 2016; Paton, Anderson, Becker & Petersen, 2015; Raby, Macabuag, Pomonis, Wilkinson & Rossetto, 2015; Contreras, Blaschke, Kienberger & Zei, 2014; Smith et al., 2012). It can be argued that what makes earthquakes so lethal, among other things, is that they cannot be predicted (Kagan, 1997; Knopoff, 1996; Jones, 1996; Sykes, 1996); people just do not know which day and at what time an earthquake can occur. This very fact has been demonstrated just recently when an earthquake of 6.2 of Magnitude hit the city of Amatrice, Central part of Italy, in August 2016. The quake hit the city at 03:36 a.m. It is believed that 292 people were killed and hundreds left injured (Withnall, 2016; Sims, 2016).

On the other hand, disasters triggered by earthquakes have been the subject of considerable research, and from different perspectives (Pan, 2012, Lindell, 1994, Drabeck, 1986, Kahnemen & Tversky, 1973, Higgins & Bargh, 1987, Santos-Reyes & Beard, 2011; Santos-Reyes, Gouzeva, Santos-Reyes, 2014a,b; Santos-Reyes, Santos-Reyes, Gouzeva & Velazquez-Martinez, 2016; Ronan, Johnston, Daly & Fairley, 2001; Paton, et al., 2015; Knopoff, 1996;

Jones, 1996, to mention just a few). For example, the topic associated with seismic risk perception has been extensively studied in an attempt to understand how individuals perceive risk (Slovic, Fishhoff, Lichtenstein, 1981; Slovic, 1987, Sparf & Ohman 2014, Pan, 2012, Lindell, 1994, Drabeck, 1986, Kahnemen & Tversky, 1973, Higgins & Bargh, 1987, Santos-Reyes, et al. 2014, Ronan, Johnston, Daly, & Fairley, 2001; Slovic, 1987).

Another important issues that has been investigated is that associated with the role of the mass media, 'information communication technologies' (ICTs), etc, in relation to, disasters, in general, and earthquakes, in particular (Ghassabi1 & Farashbandi, 2015; Weingart, 1998, 2002; Varughese, 2011; Chengc, Mitomoa, Otsukab & Jeone, 2015; Uchida, Kanagawa, Takenishi, Harada, Okawa & Yabuno, 2015). For example, Varughese (2011) contends that science and the media are two important social institutions that shape contemporary social processes. Further, Weingart (1998) argues that "With the growing importance of the media in shaping public opinion, conscience and perception on the one hand and a growing dependence of science on scarce resources and thus on public acceptance on the other, science [becomes] increasingly media oriented." This raises the question as to what should be the attitudes of the general public towards the media and in relation to earthquake disasters. However, what is clear is that the ICTs, the media have a crucial role in linking the general public with science.

Knowledge and education are recognized as key components of seismic disaster risk management. It has been argued that one of the most important lessons to emerge from earthquake disasters worldwide is that education and knowledge have the power to save lives (United Nations Children's Fund (UNICEF), 2011). Furthermore, education and knowledge can contribute to the building of a culture of prevention, safety and resilience at all levels of the society.

The book chapter intends to address seismic risk perception. Also, it presents some results associated with knowledge and education on seismic risk. Furthermore, it covers superficially some issues associated with the mass media. The sample population under study is that related to Preparatory school students in Mexico City. Overall, the results presented here are intended to serve as the basis for a more in-depth analysis, for example, of mass media, social networks, related to earthquakes, in the context of the Mexican culture. The paper gives an account of an ongoing research project.

Seismic Zones of Mexico City

Mexico City is located in a highly seismic region of the country. The City has been divided into three seismic zones as shown in Table 1 (Servicio Sismológico Nacional (SSN), 2016).

Table 1. Seismic zones of Mexico City

	School	Home	Hospital	TV	Radio	Internet	Social networks	Other
Mean	3.27	2.89	2.68	3.14	2.80	2.96	2.75	1.16
Standard deviation (SD)	0.891	1.059	1.099	0.993	1.132	1.104	1.170	0.663
Variance	0.793	1.121	1.209	0.986	1.280	1.219	1.368	0.440
N	1482	1481	1482	1482	1482	1482	1482	1482

Materials and Methods

A cross-sectional survey has been conducted; the survey employed a questionnaire of 57 questions. The results presented here include only those associated with education, knowledge and some aspects of seismic risk.

The target population, as mentioned in the introduction section, included students from six preparatory schools (PS) located within the three seismic zones described in the previous section (Table 1). The sample size of the population considered in the analysis was N = 1489. The application of the questionnaire survey took about 1.5 - 2.0 hrs to complete. Also, it should be mentioned that the first version of the questionnaire was pre-tested and administered in October 2015 by the authors.

Results and Discussion

Hazard, Risk and Natural Disaster Knowledge Sources

As mentioned in the introduction section, knowledge and education on seismic risk is an important aspect that should be a key component in any 'disaster management system' (see, for example, Santos-Reyes & Beard, 2011; UNICEF, 2010). Given this, this section presents some preliminary results regarding the sources of knowledge in relation to the terms 'Hazard', 'Risk', and 'Natural disaster'.

A series of questions were included in the questionnaire aiming at knowing the sources of knowledge; for example, a question was intended to assess to a certain degree whether the participants have heard of the following words: 'hazard', 'risk' and 'natural disaster'. However, it should be pointed out that the participants did not need to answer, for example, what the definitions of each of these were. It has been found that 99.6% (N = 1489) of the participants have heard of the word 'hazard' (Mean = 2.00, SD = 0.063), 99.5% related to 'Risk' (Mean = 2.00; SD = 0.069), and 99.2% (N = 1435) have heard of 'Natural disaster' (Mean = 2.00, SD = 0.118). Overall, the results show that the overwhelming majority of the students effectively are well aware of these terms.

Table 2 shows the results when the students were asked from which sources they heard about the term 'Risk'. The questionnaire included the following categories: "School," "Home," "Hospital," "TV," "Radio," "Internet," "Social networks," and "Other." It can be seen that the categories resulting with the highest scores were the following, and in this particular order: "School" (Mean = 3.27, SD = 0.891), "TV" (Mean = 3.14, SD = 0.993), and at "Home" (Mean = 2.89, SD = 1.059).

Table 2. Awareness of seismic hazard in the region where the children live

Seismic Zone	Description
Hill (or Zone-1)	The amplification of seismic waves is reduced and the movement is of short duration in this particular zone.
Intermediate (or Zone-2)	The zone presents characteristics intermediate between Zones 1 and 3.
Lake (or Zone-3)	This the zone where formerly two lakes were located in ancient Mexico City. Overall, the type of soil is very soft and compressible because of the lake deposits with high water content, which favours amplification of the seismic waves.

Figure 1 shows the distribution of the students responses regarding all the sources considered in the analysis for the term 'hazard' and 'risk'. For example, 24% (355/1482) of the students learn the word 'Risk' at school.

Similarly, Table 3 shows the results when the students were asked where they learned about 'Natural disaster' (as mentioned in previous section, the students were not asked to define what a natural disaster is). It can be seen that the following three categories score the highest percentage regarding this question: “TV” (Mean = 1.87, SD = 0.334), “School” (Mean = 1.85, SD = 0.362), and in the “Internet” (Mean = 1.78, SD = 0.415). Full details about the distribution of the frequency of data regarding each of the categories is shown in Figure 2.

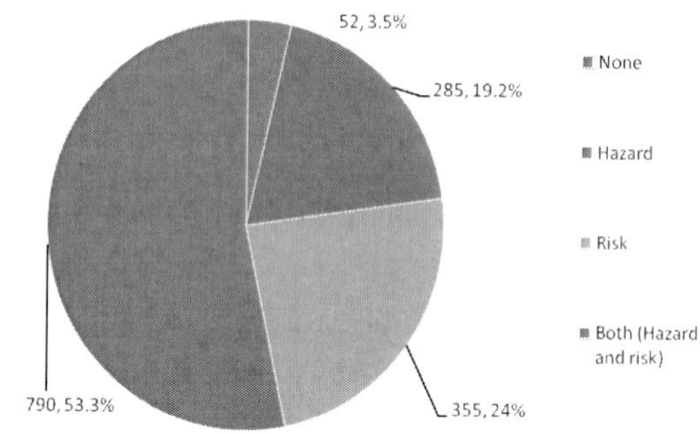

(a). School.

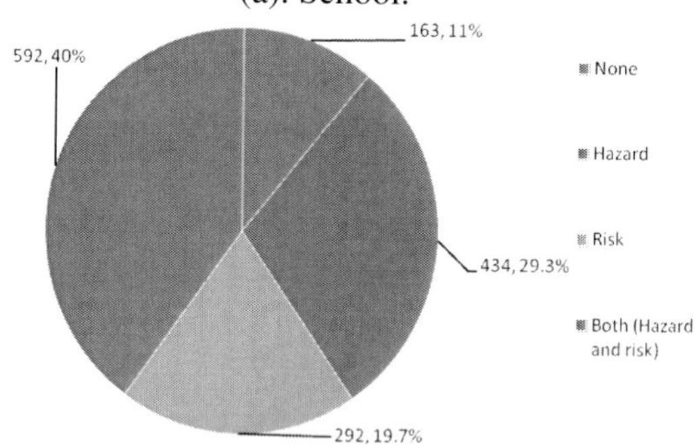

(b). House

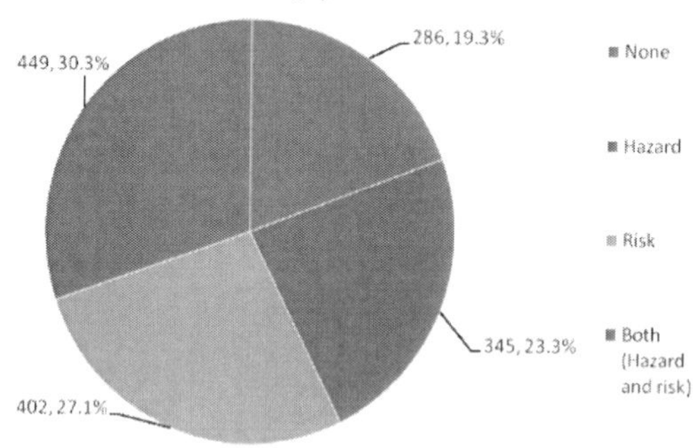

(c). Hospital

Figure 1. (Continued).

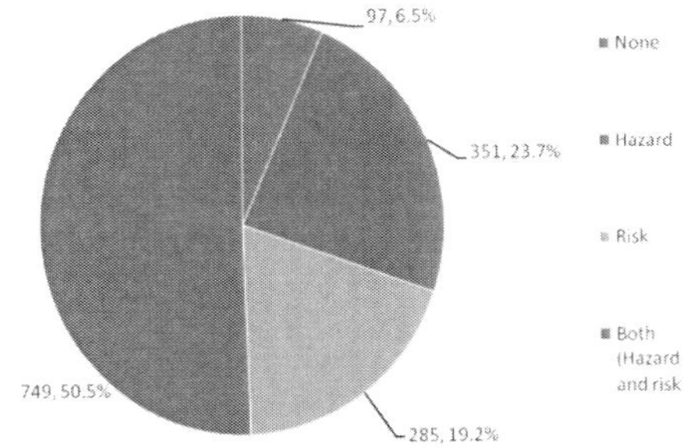

(d). TV

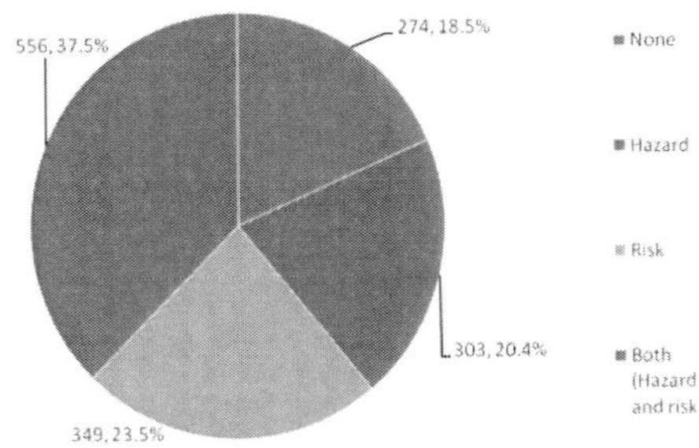

(e). Radio

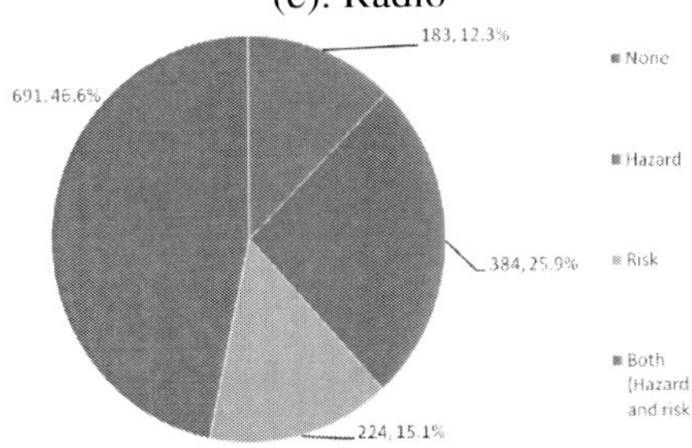

(f). Internet

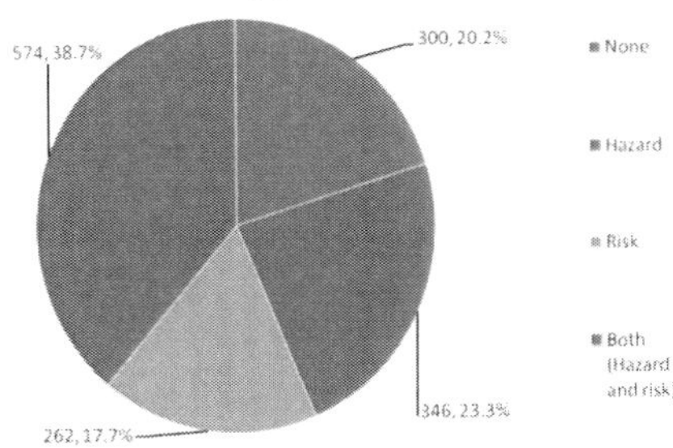

(g). Social networks.

Figure 1. Sources of information regarding ‘Hazard’ and ‘Risk’.

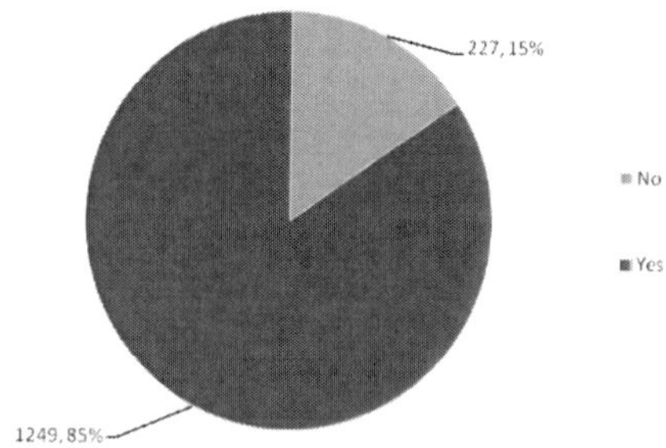

(a). School.

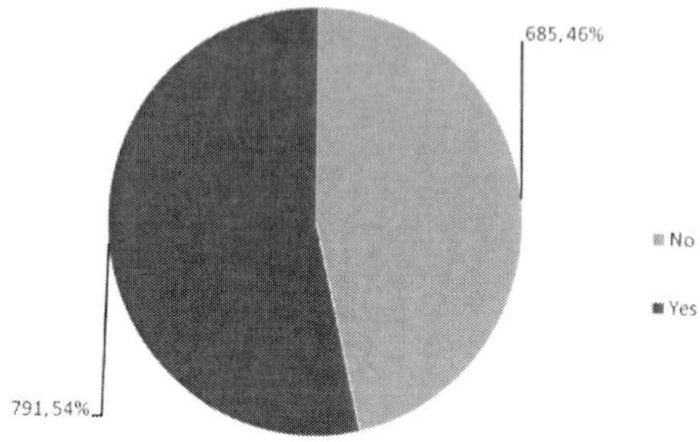

(b). House.

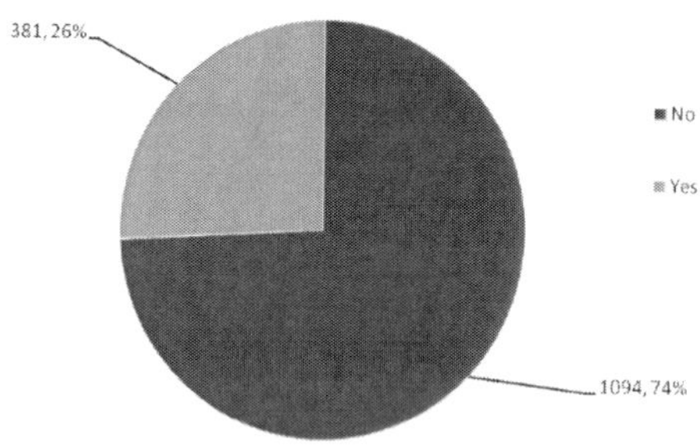

(c). Hospital.

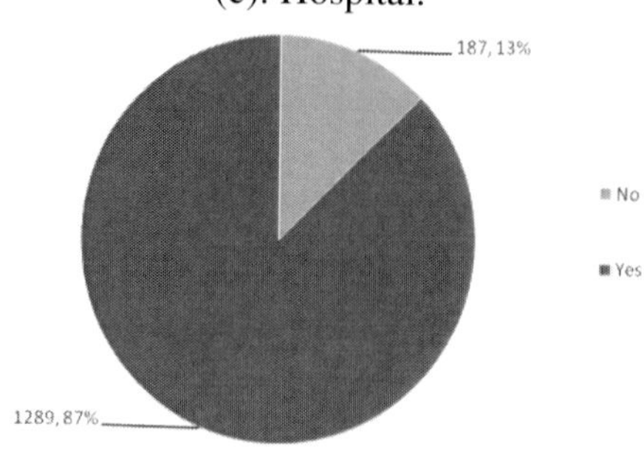

(d). TV

Figure 2. (Continued).

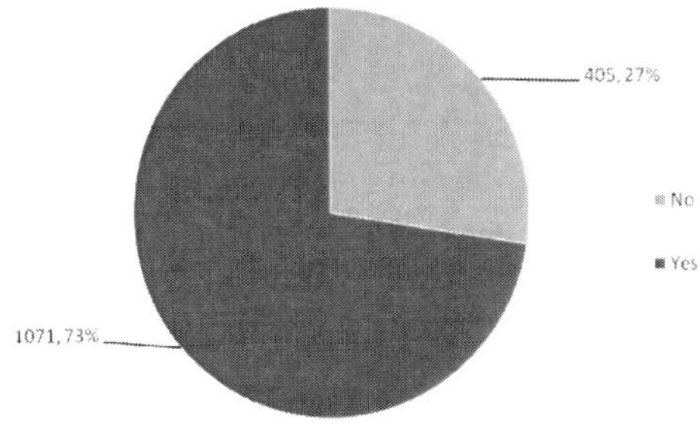

(e). Radio.

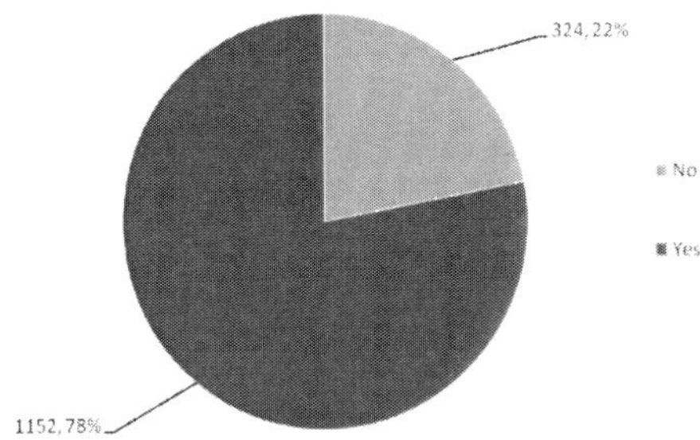

(f). Internet.

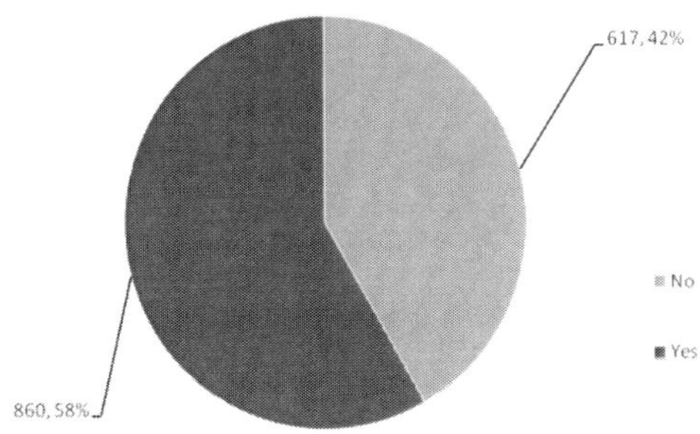

(g). Social networks.

Figure 2. Sources of information regarding the term 'Natural disaster'.

Overall, it is interesting to see how the students learnt about the term 'Natural disaster' through the mass media such as the "TV" and from the "Internet." Effectively, the modern world is characterised by the age of the 'information and communication technologies' ("Internet," "Social networks") and clearly the students are more familiar with such technologies; further, it is common now a days, for example, to see the destruction caused by disasters triggered by earthquakes in the mass media (Sims, 2016; Gassbil &

Farashbandi, 2015; Uchida, et al., 2015; Changc, et al., 2015; Varughese, 2011).

Table 3. Students' sources of knowledge related to 'Natural disaster'

	School	Home	Hospital	TV	Radio	Internet	Social networks	Other
Mean	1.85	1.54	1.26	1.87	1.73	1.78	1.58	1.04
Standard deviation (SD)	0.362	0.500	0.439	0.334	0.447	0.415	0.493	0.194
Variance	0.131	0.250	0.193	0.112	0.200	0.172	0.243	0.038
N	1477	1477	1477	1477	1477	1477	1477	1476

It was quite surprising that "Social networks" (Twiter, Facebook, Google, etc.) came fifth in relation of the information source of the term 'natural disaster'. It is believed that during the 2015 Chilean earthquake, social networks played an important role; i.e., given their feature associated with instant messaging applications Chileans could manage to communicate, request help or simply informed about the earthquake to their friends, colleagues, etc. (El Mundo, 2015).

The participants of the study were asked whether they considered that a natural disaster could affect their community. The results show that 93.2% (1377/1489, Mean=1.93, SD=0.252) responded as "Yes."

As mentioned in the previous sections, the population sample has been considered in relation to the location of the schools within the seismic zones of the Capital City. The following question has been put to the participants: "Do you know that the City is divided into three seismic zones?" The results were quite surprising for us; i.e., 89.1% (1327/1489) did not know this fact. We expected the students to do better than they did in this regard (Mean = 1.10, SD = 0.294).

The results of the variables associated with gender, location of the Preparatory school (PS) and the age are shown in Figure 3 and Tables 4 & 5, respectively.

Figure 3 shows clearly that when considering the gender of the participants, 89.9% (642/1467) of males and 91% (685/1467) of females, did not know the fact that the city is being divided into three seismic zones.

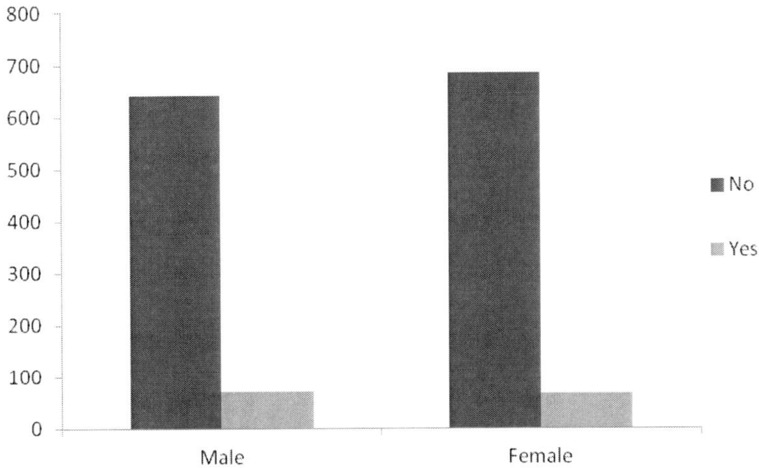

Figure 3. Students' response according the variable Gender.

Table 4, on the other hand, shows the results when considering the variable associated with the location of the PSs. The results show that the following schools rated the highest percentage of not knowing this fact and in this particular order: PS-3 (93.9%), PS-1 (93.8%), and PS-4 (91.3%)

Table 4. Results according to the variable related to the location of the PSs

Answers	Preparatory School-1 (PS-1) N(%)	Preparatory School-2 (PS-2) N(%)	Preparatory School-3 (PS-3) N(%)	Preparatory School-4 (PS-4) N(%)	Preparatory School-5 (PS-5) N(%)	Preparatory School-6 (PS-6) N(%)
No	166 (93.8)	172 (87.3)	321 (93.9)	273 (91.3)	238 (88.5)	157 (85.8)
Yes	11 (6.2)	25 (12.7)	21 (6.1)	26 (8.7)	31 (11.5)	26 (14.2)
N	117	197	342	299	269	183

Finally, when considering the age of the participants, the results show that as expected older students appear to be more knowledgeable about the seismic zones than younger students (14.9% and 4.5%, respectively), as shown in Table 5. On the other hand, it can be seen that the students from the ages of 14, 15, 17 and 18 years old scored the highest percentage of not knowing this fact.

Table 5. Results associated with the variable Age of the participants

Answers	14 Years old N(%)	15 Years old N(%)	16 Years old N(%)	17 Years old N(%)	18 Years old N(%)	19 Years old N(%)
No	42 (95.5)	295 (90.2)	451 (89.8)	359 (91.1)	140 (91.5)	40 (85.1)
Yes	2 (4.5)	32 (9.8)	51 (10.2)	35 (8.9)	13 (8.5)	7 (14.9)
N	44	327	502	394	153	47

Overall, the results presented in the section, clearly shows the lack of awareness and knowledge on seismic risk key issues, such as the seismic zones of the Capital City. It may be argued that this is important when considering all aspects associated with preparedness before an earthquake strikes. Furthermore, the results presented in the section illustrates the lack of a culture of prevention and this further contributes to the vulnerability of the population considered in the study.

Seismic Risk Perception

A question related to when an earthquake could occur in the future was put to the students. The possible answers were the following: "Never," "In your lifetime," "Within the next 10 years," "Next year," and "At any time." The results are shown in Table 6, Figures 4 & 5.

In general, when considering the variable gender, the results show that the overwhelming majority considered the occurrence of an earthquake "At any time" (Male = 82.3%; Female = 91.3%). In particular, females scored slightly higher than males in this regard (Table 6).

Similarly, when considering the age of the preparatory school students, the results show a similar trend as with the gender. That is, students from 14 to 19 years old are well aware that an earthquake effectively can happen "At any time." As expected, older students scored slightly higher than younger students in this respect, as shown in Table 6 and Figure 4.

Also, the analysis considered the variable associated with the schooling mode of the preparatory schools that participated in the study; i.e., in Mexico, in general, there are two schooling modes, the morning (08:00-14:00 hrs) and the afternoon (14:00-20:00 pm). As expected, similar results as above have been found; i.e., the occurrence of an earthquake at any time (Morning: 89.3%, Afternoon:82.9%); see Table 6.

Table 6. Summary of the results associated with the key demographic variables considered in the study

		"Never" N (%)	"In your lifetime" N (%)	"Within 10 years" N (%)	"Next year" N (%)	"Any time" N (%)	Total (N)
Gender	Male	14 (2.0)	40 (5.6)	29 (4.1)	43 (6.0)	586 (82.3)	712
	Female	5 (0.7)	20 (2.7)	15 (2.0)	25 (3.3)	685 (91.3)	750
Age (Years old)	14	0 (0.0)	2 (4.4)	1 (2.2)	2 (4.4)	40 (88.9)	45
	15	7 (2.2)	23 (7.1)	12 (3.7)	18 (5.6)	264 (81.5)	324
	16	7 (1.4)	19 (3.8)	13 (2.6)	20 (4.0)	437 (88.1)	496
	17	4 (1.0)	12 (3.0)	12 (3.0)	17 (4.3)	351 (88.6)	396
	18	1 (0.6)	3 (1.9)	4 (2.6)	10 (6.5)	136 (88.3)	154
	19	0 (0.0)	1 (2.1)	2 (4.3)	1 (2.1)	43 (91.5)	47
Preparatory school	Preparatory School-1 (PS-1)	2 (1.2)	8 (4.7)	4 (2.3)	6 (3.5)	151 (88.3)	(171)
	Preparatory School-2 (PS-2)	4 (2.0)	2 (1.0)	2 (1.0)	5 (2.5)	187 (93.5)	(200)
	Preparatory School-3 (PS-3)	6 (1.7)	16 (4.7)	16 (4.7)	20 (5.8)	285 (83.1)	(343)
	Preparatory School-4 (PS-4)	0 (0.0)	11 (3.7)	6 (2.0)	9 (3.0)	273 (91.3)	(299)
	Preparatory School-5 (PS-5)	6 (2.2)	14 (5.2)	11 (4.1)	21 (7.8)	216 (80.6)	(268)
	Preparatory School-6 (PS-6)	1 (0.6)	9 (5.0)	5 (2.8)	7 (3.9)	159 (87.8)	(181)
Schooling mode	Morning	10 (1.1)	27 (2.9)	20 (2.2)	41 (4.5)	820 (89.3)	(918)
	Afternoon	9 (1.7)	33 (6.1)	24 (4.4)	27 (5.0)	451 (82.9)	(544)

Finally, when considering the variable related to the school location, the results are shown in Table 6 and Figure 5. It can be seen that in general, all schools (i.e., the participants of the study) considered that an earthquake can occur at any time. In particular, the PS-2 (which is located in what is called the

Transition Zone, see Table 1) with 93.5% (187/200), followed by PS-4 (91.3%, 273/299) which is located in the Lake Zone or Zone-3), and finally PS-6 (87.8%, 159/181, located in the so called Hill zone or Zone-1). See Figure 5.

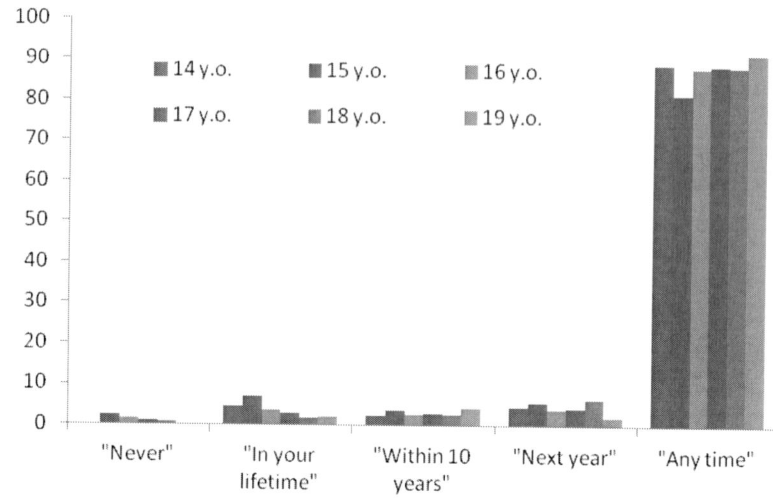

Figure 4. Students' risk perception-variable Age.

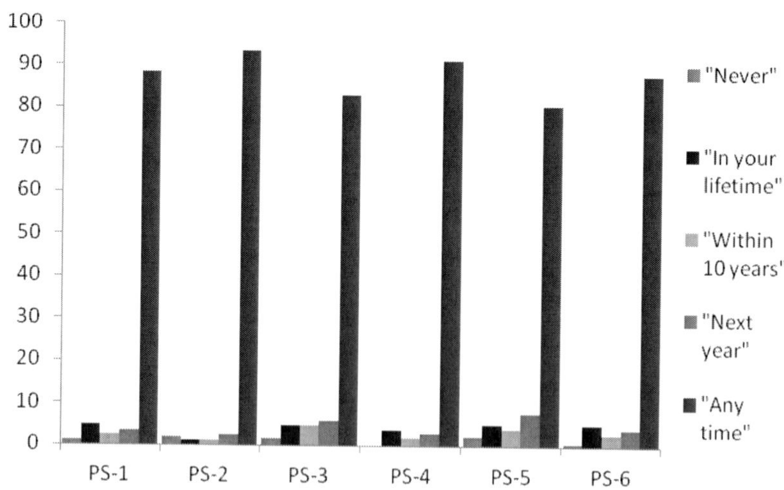

Figure 5. Students' risk perception-variable related to the location of the Preparatory schools (PSs).

Overall, the results presented in this section addressed a very important issue associated with the seismic risk perception of the students as a future event. Interestingly, the participants are well aware that an earthquake could strike "At any time" and this may be welcomed; however, the participants are not well prepared to such events (see the previous section).

As for the variable associated with the location of the Preparatory schools (PSs), we expected the ones located in the Zone-3 (or Lake) to do better that they actually did. That is, this particular zone has a very high degree of vulnerability to earthquakes, as demonstrated by the 1985 earthquake that hit the Capital City (Gutierrez, Santoyo, Quaas, Ordaz & Singh, 2008).

Seismic Hazard Knowledge Sources

It is clear that education and knowledge for disaster risk reduction is essential to mitigate the impact of earthquakes (Nathe, 2000; UNICEF, 2010). It has been argued that the "use of knowledge, innovation and education to build a culture of safety and resilience at all levels" (UNICEF, 2010), could effectively contribute to achieve the above. Furthermore, education and knowledge could contribute to a drastic shift in "mentalities and perceptions as well as a behavioral change towards a more proactive preventative approach to disasters" (UNICEF, 2010).

Given the above, we included a series of question aiming at gaining an understanding of the ways the participants considered the best ways to learn more about seismic risk. The following categories were considered, as the sources of education: "Books, "Internet," "School," "Civil Protection," "Radio & TV", and "Other". Table 7 and Figures 6, 7 & 8 show the results of the analysis.

Figure 6 shows the results of the preferences of the participants of the study when considering the sample population as a whole. It can be seen that 44% considered "Civil Protection" as the best way learn more about seismic risk. This was followed by "School" (22%), "Internet" (20%), and special programmes on "Radio and TV" came fourth (10%).

On the other hand, when considering the variable associated with gender, the results show that males have the following preferences and in this order:" Civil Protection," "Internet," and "School"; females, on the other hand, show a tendency similar to that of the general population sampled (Table 7 and Figure 7).

Table 7. Summary of the results related to the variable Gender and Age of the participants of the study

		"Books" N (%)	"Internet" N (%)	"School" N (%)	"Civil protection" N (%)	"Radio & TV" N (%)	"Other" N (%)	Total (N)
Gender	Male	30 (5.0)	163 (27.0)	125 (20.7)	239 (39.6)	47 (7.8)	0 (0.0)	(604)
	Female	21 (3.3)	86 (13.4)	149 (23.2)	310 (48.4)	70 (10.9)	3 (0.5)	(641)
Age (Years old)	14	6 (15.0)	11 (27.5)	8 (20.0)	15 (37.5)	0 (0.0)	0 (0.0)	(40)
	15	12 (4.4)	59 (21.5)	53 (19.3)	124 (45.1)	25 (9.1)	0 (0.0)	(275)
	16	16 (3.8)	80 (19.1)	92 (22.0)	195 (46.5)	34 (8.1)	2 (0.5)	(419)
	17	14 (4.2)	66 (20.0)	73 (22.1)	139 (42.1)	37 (11.1)	1 (0.3)	(330)
	18	3 (2.1)	24 (16.9)	42 (29.6)	59 (41.5)	14 (9.9)	0 (0.0)	(142)
	19	0 (0.0)	0 (0.0)	6 (15.4)	17 (43.6)	7 (17.9)	0 (0.0)	(39)

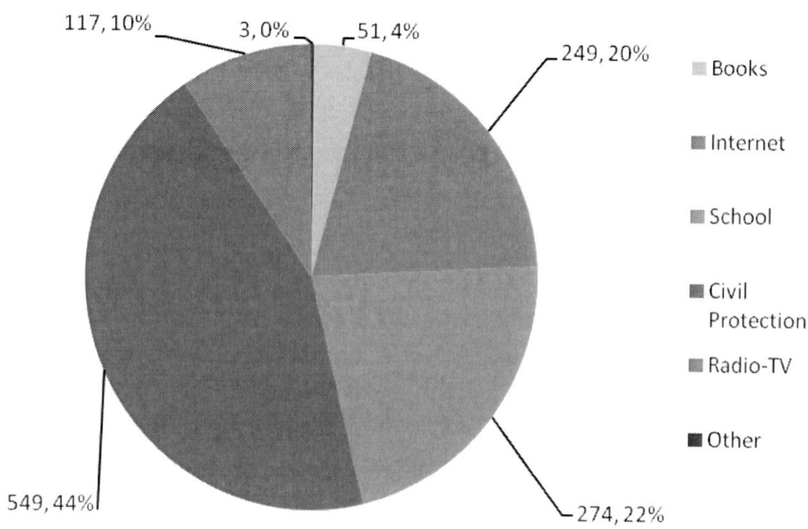

Figure 6. Categories as a source of seismic education.

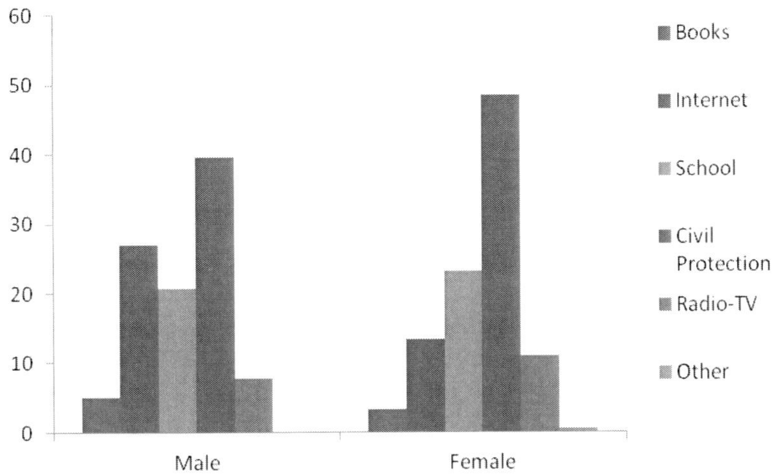

Figure 7. Preferred sources of education on seismic risk-variable Gender.

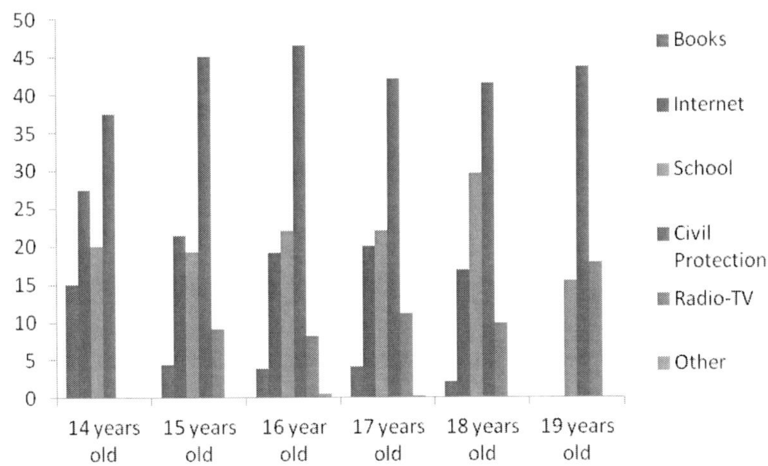

Figure 8. Preferred sources of education on seismic risk-variable Age.

When considering the age of the participants of the study, the results show that the preferences are similar to those mentioned above; i.e., for the case of students from the ages between 16-18 years old (i.e., "Civil Protection"-"School"-"Internet"). Students from the ages of 14 and 15 years old, on the other hand, their preferences were the following: "Civil protection"-"Internet"-"School." Finally, older students (i.e., 19 years old), preferred "Radio & TV" as the second option after "Civil Protection," as their preferred way to learn more about seismic risk. See Figure 8.

In general, it can be argued that one of the most important lessons to emerge from the series of devastating disasters caused by earthquakes worldwide is that education and knowledge have the power to save lives (UNICEF, 2010). For example, it is believed that in a recent study, it has been estimated that roughly "1 billion children aged 1-14 live in countries with high seismic risk, which puts several hundred million children at risk while they are attending schools" (UNICEF, 2010). Furthermore, education and knowledge for disaster seismic risk reduction should foster the creation of a resilient culture; as UNICEF (2010) put it: "Children, as "tomorrow's leaders" and key agents for change, are recognized as the primary targets of these efforts."

CONCLUSION

The main results are the following:

- Overall, the results show that the students are well aware of the words 'Hazard', 'Risk', and 'Natural disaster'; i.e.,: 'Hazard' (99.6%, Mean = 2.00, SD = 0.063), 'Risk' (99.5%, Mean =2.00, SD = 0.069), and 'Natural disaster' (99.2%, Mean = 2.00, SD = 0.118).
- The analysis showed that the main source of information associated with the term 'Risk' (Mean = 3.27, SD = 0.891) was the "School". The term 'Natural disaster', on the other hand, has been found to be "TV" (Mean = 1.87, SD = 0.334), followed by "School" (Mean = 1.85, SD = 0.362). It was found that "Social networks" came fifth as the source of information related to 'Natural disaster'.
- Regarding the participants' knowledge on the three seismic zones of the Capital City, the overwhelming majority of them did not know this fact (89.1%, Mean = 1.10, SD = 0.294).
- Preparatory school students perceived that an earthquake could occur "At any time" (86.9%, Mean = 4.72, SD = 0.816).
- Preparatory school students prefer to learn more on seismic risk by the following sources and in this particular order: "Civil protection" (44%), "School" (22%), and "Internet" (20%).
- In general, the participants of the study showed a lack of knowledge on seismic risk (see c) above), even though, they are well aware that an earthquake could happen at any time (see d) above).

ACKNOWLEDGMENTS

This project was funded under the following grants: CONACYT, No-248219 & SIP-IPN: No-20161496.

REFERENCES

Beck, E., André-Poyaud, I., Paule-Annick, D., Chardonnel, S., & Lutoff, C. (2012). Risk perception and social vulnerability to earthquakes in Grenoble (French Alps), *Journal of Risk Research, 15*(10), 1245-1260.

Blaikie, P., Cannon, T., Davis, I., & Wisner, B. (1994). At risk: natural hazards, people's vulnerability, and disasters. London. Routledge.

Chengc, J.W., Mitomoa, H., Otsukab, T., & Jeonc, S.Y. (2015). The effects of ICT and mass media in post-disaster recovery –A two model case study of the Great East Japan Earthquake. *Telecommunications Policy*, *39*(6), 515–532.

Contreras, D., Blaschke, T., Kienberger, S., & Zeil, P. (2014). Myths and realities about the recovery of L'Aquila after the earthquake. *International Journal of Disaster Risk Reduction*, *8*, 125-142.

Drabeck, T. (1986). *Human System Responses to Disasters: An Inventory of Sociological Findings.*, New York. Springer-Verlag.

El Mundo. (2015). *"Por qué las redes sociales fueron claves durante el terremoto de Chile"* Retrievde from http://www.elmundo.es/internacional/2015/09/18/55fbeee946163f81488b457f.html. [*Why the social networks were the key during the Chilean earthquake?*].

Ghassabi1, F., & Farashbandi, F.Z. (2015). The role of media in crisis management: A case study of Azarbayejan earthquake. *International Journal of Health System Disaster Management*, *3*(2), 95-102.

Gutierrez, C., Santoyo, M., Quaas, R., Ordaz, M., & Singh., S.K. (2008). *Fascículo Sismos*, 5a. Edición, Mexico. CENAPRED. [*Earthquakes Booklet*, 5th. Edition, Mexico. CENAPRED].

Higgins, E.T., & Bargh, J. A. (1987). Social cognition and social perception. *Annual Review of Psychology, 38,* 369-425.

Jones, L.M. (1996). Earthquake prediction: The interaction of public policy and science. *Proc. Natl. Acad. Science, 93*, 3721-3725.

Kagan, Y.Y. (1997). Assessment of schemes for earthquake prediction: Are earthquakes predictable? *Geophysical Journal International, 13*(3), 505-525.

Kahnemen & Tversky. (1973). On the psychology of prediction. *Psychological Review, 80*(4), 237-251.

Knopoff, L. (1996). Earthquake prediction: The scientific challenge. *Proc. Natl. Acad. Science,* 93, 3719-3720.

Lindell, M. K. (1994). Perceived Characteristics of Environmental Hazards. *International Journal of Mass Emergencies and Disasters, 12*(3), 303-326.

Nathe, S. 2000. Public Education for Earthquake Hazards. *Natural Hazards Review*, *1*(4), 191-196.

Pan, A. (2012). A Study on Residents' Risk Perception in Abrupt Geological Hazard. *Journal of Risk Analysis and Crisis Response, 2*(1), 44-55.

Paton, D., Anderson, E.., Becker, J., & Petersen, J. (2015). Developing a comprehensive model of hazard preparedness: Lessons from the Christchurch earthquake. *International Journal of Disaster Risk Reduction, 14*(1), 37-45.

Raby, A., Macabuag, J., Pomonis, A., Wilkinson, S., & Rossetto, T. (2015). Implications of the 2011 Great East Japan Tsunami on sea defence design. *International Journal of Disaster Risk Reduction, 14*(4), 332-346.

Ronan, K.R., Johnston, D.M., Daly, M., & Fairley, R. (2001). School children's risk perception and preparedness: a hazard education survey. *The Australasian Journal of Disaster and Trauma Studies, 1*, 1-35.

Santos-Reyes J. & Beard A.N. (2011). Applying the SDMS model to the analysis of the Tabasco flood disaster in Mexico, *Human and Ecological Risk Assessment; An International Journal, 17*(3), 646-677.

Santos-Reyes, J., Gouzeva, T, & Santos-Reyes, G. (2014a). Earthquake risk perception and communication: A review of empirical research. *Disaster Advances, 7*(7), 77-87.

Santos-Reyes, J., Gouzeva, T., & Santos-Reyes, G. (2014b). Earthquake risk perception and Mexico City's public safety. *Procedia Engineering, 84*, 662 – 671.

Santos-Reyes, J., Santos-Reyes, G., Gouzeva, T., & Velazquez-Martinez, D. (2016). Schoolchildren's earthquake knowledge, preparedness and risk perception of a seismic prone region of Mexico. *Human and Ecological Risk Assessment: An International Journal*. In press.

Sims, A. (2016). *The quake struck in the early hours of the morning when most residents were asleep*. The Independent, Retrieved from http://www.independent.co.uk/news/world/europe/italyearthquaketodaydeathtolllatestnewsamatriceaccumolimagnitudedevastationa7208046.html

Slovic, P. (1987). Perception of risk. *Science,* 236, 280-285.

Slovic, P. Fishhoff, B., & Lichtenstein, S. (1981). *Perceived risk: psychological factors and social implications.* Proceedings of the Royal Society of London, A376, 17-34.

Smith, R.M., Dyer, G.S., Antonageli, K., Arredondo, N., Bedilion, H., Dalal, A., Deveny, G.M., Joseph, G., Lauria, D., Lockhart, S.H., Lucien, S., Marsh, S., Rogers, S.O., Salzarulo, H., Shah, S., Toussaint, R.J., & Wagoner, J. (2012). Disaster triage after the Haitian earthquake. *Injury, 43*(11), 1811-1815.

Sparf, J., & Öhman, S. 2014. On Risk and Disability – Investigating the Influence of Disability and Social Capital on the Perception and Digital Communication of risk. *Journal of Risk Analysis and Crisis Response* 4(1): 20-33.

Sykes, L. (1996). Intermediate and long-term earthquake prediction. *Proc. Natl. Acad. Science,* 93, 3732-3739.

Uchida, Y., Kanagawa, C., Takenishi, A., Harada, A., Okawa, K., & Yabuno, H. (2015). How Did the Media Report on the Great East Japan Earthquake? Objectivity and Emotionality Seeking in Japanese Media Coverage. *PLoS One. 0*(5): e0125966. doi: 10.1371/journal.pone.0125966.

UNICEF. (2010). *Using education to reduce risk- An overview of disaster risk reduction and how it is promoted through education in the SEE/CIS region.* UNICEF Regional Office for CEE/CIS Education, Geneva 10 Switzerland.

Varughese, S.S. (2011). Media and Science in Disaster Contexts: Deliberations on Earthquakes in the Regional Press in Kerala, India. *Spontaneous Generations: A Journal for the History and Philosophy of Science*, *5*(1), 36-43.

Weingart, P. 1998. Science and the Media. *Research Policy, 27*(8), 869-79.

Weingart, P. (2002). The Loss of Distance: Science in Transition. In *Science, History, and Socialism: A Tribute to Evere Mendelsohn,* eds. Garland E. Allen and Roy M. MacLeod, 167-84. Dordrecht: Kluwer Academic Publishers.

Withnall, A. (2016). Italy earthquake: Extraordinary stories of survival emerge as death toll from 6.2 magnitude quake rises to 241. Retrieved from http://www.independent.co.uk/news/world/europe/italyearthquakelatestnewsvideoamatricemapdeathtollmagnitudesurvivorsstoriesa7209591.html.

In: Earthquakes
Editor: Wayne Coleman

ISBN: 978-1-53610-342-7

Chapter 6

SEISMIC RISK PERCEPTION AND PREPAREDNESS OF PRIMARY SCHOOLCHILDREN: THE CASE OF HUAJUAPAN, OAXACA, MEXICO

Jaime Santos-Reyes**, *Galdino Santos-Reyes* and *Tatiana Gouzeva
Grupo de Investigación SARACS, ESIME, ZAC.,
Instituto Politecnico Nacional, México

ABSTRACT

The chapter presents results regarding urban and semi-rural schoolchildren's seismic risk perception, awareness and preparedness of a highly seismic region, in the state of Oaxaca, Mexico. The sample of the study was N = 819 (Urban sample size = 522; Semi-rural sample size = 297). The main findings of the cross-sectional study have been the following: a) overall, urban schoolchildren demonstrated a generalized awareness of the seismic area where they live; however, this was not the case for semi-rural schoolchildren; educational campaigns may be needed to further enhance seismic awareness of both urban and semi-rural areas of the region; b) the two hazards associated with earthquakes and crime

* Corresponding author: Tel.: +52 55 57296000, Ext. 54663; Fax: +52 55 57296000, Ext. 54588. Email address: jsantosr@ipn.mx; jrsantosr@hotmail.com (J. Santos-Reyes).

have been perceived to represent the two major threats to them at school and at home, and for both urban and semi-rural schoolchildren; c) schoolchildren emotions are not consistent when asked whether they would be able to control themselves during an earthquake; d) seismic risk communication takes place at school and at home; this may be regarded as encouraging in order to further educate children on seismic risk; however, more is needed to better educate them on important aspects regarding what to do before, during and after an earthquake and this may well prove to be an effective way to share their knowledge with their parents (most of them are not well educated) and contribute to built a resilient community to earthquakes; e) the preliminary findings of the study are consistent with has been found in similar studies; i.e., "children were able to identify higher frequency events at a higher frequency and lower frequency events at a relatively lower frequency."

Keywords: school children, earthquakes, Mexico, seismic risk perception

INTRODUCTION

In recent years, a number of disasters caused by earthquakes have demonstrated the vulnerability of both the developing and the developed world (Paton, Anderson, Becker & Petersen, 2015; Raby, Macabuag, Pomonis, Wilkinson & Rossetto, 2015; Contreras, Blaschke, Kienberger & Zei, 2014; Smith, et al., 2012; Sims, 2016). Even though, vulnerability is being associated with the social, political and economic situation of the community at disaster risk. Vulnerability has been defined as "the characteristics of a person or group in terms of their capacity to anticipate, cope with, resist and recover from the impact of a natural hazard" (Blaikie, Cannon, Davis & Wisner, 1994).

Over the last thirty years, scholars have carried out research in an attempt to understand, among other things, how individuals perceive risk. Furthermore, to attempt to explain the differences in individual risk perception, and consequently to develop approaches for assessing the viewpoints that individuals have about risk (Slovic, Fishhoff, Lichtenstein,1981; Slovic, 1987, Sparf & Ohman 2014, Pan 2012, Lindell, 1994, Drabeck, 1996, Kahnemen & Tversky, 1973, Higgins & Bargh, 1987, Santos-Reyes, Gouzeva, Santos-Reyes, 2014a,b, 2014; Ronan, Johnston, Daly & Fairley, 2001). Higgins & Bargh (1987) argue that individuals tend not to reason about risk by weighing and combining available evidence in a rational way but employ a number of mental strategies, which sometimes yield reasonable judgments and sometimes lead to severe systematic errors. Other scholars such as Slovic (1987) contends that communities' perception of risk is often found biased to overestimating small probability events and underestimating large ones; these perceptions, whether based on fact or not, have a moderating effect on the types of protective measures or preparedness behaviour that individuals choose to undertake about risk.

Overall, it may be argued that there is a lack of research regarding children's seismic risk perception, in general, and schoolchildren, in particular. The research reported in the literature is in the form of children's reactions to the occurrence of a disaster (Ronan & Johnston, 1996). The general findings are that children's reactions to hazards are based on a combination of factors that include: (a) direct exposure to the hazard combined with the perception of increased physical risk, (b) pre-existing characteristics, etc. (Ronan & Johnston, 1996, Ronan et al. 2001). Also, it may be argued that research on this area is effectively adult-based as discussed above. In addition, findings with adult populations may have a beneficial influence on theoretical perspectives related to children-based risk perception, preparedness. However, children are not adults. More importantly, they often do not have the same level of independence of action that can allow for consistent and adult based risk mitigation; i.e., preparedness at home or school setting is more often a function of adult activities. For example, designing and how to handle warning systems, evacuation plans, provision of shelter, etc. (Ronan et al. 2001).

The book chapter intends to address this gap by assessing risk perception, awareness and preparedness of schoolchildren of a highly seismic region of, Huajuapan, Oaxaca, Mexico. The paper gives an account of an ongoing research project.

SEISMICITY OF HUAJUAPAN, OAXACA, MEXICO

Mexico is one of the highly seismic countries in the world. It is located within five tectonic plates. According to historical data, seismic events with greater impact than have occurred in the country, have originated along the Pacific coast in the subduction zone plate Cocos and Rivera under the North American plate. It is believed that in this area the angle of dip of the Cocos plate is approximately 14°, the relative velocity with respect to North America plate is about 6.4 cm/year and its contact zone is about 16 km depth along a strip of 80 km (Lermo et al. 1991).

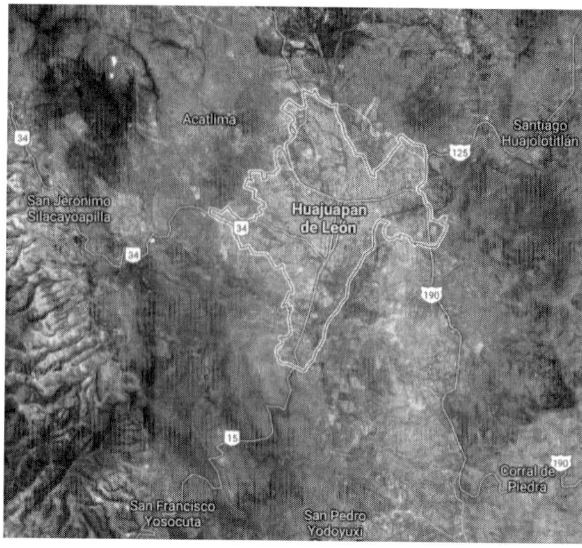

Figure 1. Location of the city of Huajuapan (Oaxaca) and the surrounding communities where the study was conducted.

The state of Oaxaca, on the other hand, is located in an area of high seismicity of Mexico and it has been divided into eight seismic zones. The city of Huajuapan (i.e., the urban area in our study) and the surrounding communities (i.e., the semi-rural area) are located within the seismic zone-2 (Figure 1). Overall, the tremors of this particular seismic zone primarily affect the southern state of Puebla, the northwest of Oaxaca and north-eastern of the state of Guerrero (these are not shown here). The zone-2, is being characterized, in relation to the epicentre, that the intensity of an earthquake is stronger in the south than in the north of the zone. Within this zone there have been two major earthquakes; i.e., the one that occurred on July 19, 1882, which was felt in much of the country and the earthquake of October 24, 1980 (M = 7) (Nuñez-Cornú & Ponce, 1989).

MATERIALS AND METHODS

A survey questionnaire has been designed in order to collect data and to assess, among other things, seismic risk perception, the level of awareness and preparedness of schoolchildren in both urban and semi-rural locations of the city of Huajuapan, Oaxaca, México (Figure 1). The first version of the questionnaire was pre-tested and administered in October 2014. In addition, it took about 50 min to complete the survey (also, it should be mentioned that two psychologists assisted in the process of administering it).

The cross-sectional study has been conducted to a sample size of N = 819 schoolchildren (Table 1); i.e., eleven primary schools agreed to take part in the survey and represented a range of demographics. Six schools were from semi-rural areas (N = 297) and five from urban areas (N = 522) of the region of 'Huajuapan' (Figure 1).

RESULTS OF THE ANALYSIS

General Knowledge

In order to assess schoolchildren's general knowledge on seismic risk, a series of questions have been included in the questionnaire; for example, the following: "Do you know that you live in a highly seismic region?" Schoolchildren were asked to rate whether they know that they live in a seismic prone region, by using a 2-point scale: 0 = "No", 1 = "Yes." The results are shown in Table 2.

Table 1. Demographic characteristics of the schoolchildren considered in the study

Variables				N	%
Gender	Urban	Male		250	47.9
		Female		272	52.1
			Subtotal	522	100.0
	Semi-rural	Male		152	51.2
		Female		145	48.8
			Subtotal	297	100.0
Age	Urban		7-8.9	43	8.8
		Age	9-10.9	220	42.2
		(years old)	11-12.9	233	44.6
			13-15	26	4.9
	Semi-rural		7-8.9	32	10.7
		Age	9-10.9	108	36.3
		(years old)	11-12.9	138	46.5
			13-15	19	6.3
Education	Primary School		Total	819	100.0

Table 2. Awareness of seismic hazard in the region where the children live

	Possible Answer	N	%	Mean (SD)
Urban schoolchildren	"Yes"	313	63.5	0.6349 (0.4819)
	"No"	180	36.5	
	Total	493	100.0	
Semi-rural schoolchildren	"Yes"	125	44.8	0.4480 (0.4981)
	"No"	154	55.2	
	Total	279	100.0	

Table 3. Children's responses regarding their knowledge on seismic risk

	Possible answers	N	%
Urban schoolchildren	"Not good"	42	8.3
	"Little"	147	29.0
	"Neutral"	149	29.4
	"Good"	121	23.9
	"Very good"	48	9.5

It can be seen that the majority of schoolchildren from the urban area recognised that effectively they live in a highly seismic region (Mean = 0.6349; SD = 0.4819; 63%, 313/493); on the other hand, 1 in 3 participants did not endorse this fact.

The same question was put to the semi-rural schoolchildren, the results show a mean of 0.4480 (SD = 0.4981). This means that semi-rural schoolchildren that participated in the study do not recognised that they live in a seismic prone region. The frequency data shown in Table 2 shows this trend.

A question has been put to the schoolchildren about the degree of knowledge they have in relation to earthquakes in general; e.g., "What degree of knowledge do you have about earthquakes?" Children were asked to rate their answer to a 5-point Likert scale: 1 = "Not good," 2 = "Little," 3 = "Neutral," 4 = "Good," and 5 = "Very good." The calculated mean was 2.97 (SD = 1.113) which may be interpreted as an average that urban schoolchildren have little knowledge on seismic risk. The frequency data shown in Table 3 supports this trend. Also, Table 3 shows that 1 in 5 urban schoolchildren agree that they have a good knowledge on earthquakes and 1 in 10 endorse that they possess a very good knowledge on seismic risk. Similar results were obtained when asking the same question to semi-rural schoolchildren; i.e., they endorse little knowledge on earthquakes (Mean = 2.8, SD = 1.130).

Table 4. Participants' responses regarding the cause of the occurrence of an earthquake

	Possible answers	7 Years old (yo)	8 yo	9 yo	10 yo	11 yo	12 yo	13 yo	14 yo	15 yo
Urban schoolchildren	"God's punishment"	0	4	6	5	3	5	2	0	0
	"Slip of tectonic plates"	0	4	44	74	100	96	14	5	0
	"Excess of heat"	0	3	2	0	0	2	1	0	0
	"After an hurricane"	0	5	8	3	3	1	0	0	0
	"I don´t know"	0	24	46	21	7	9	1	2	0
Semi-rural schoolchildren	"God's punishment"	0	2	4	5	2	1	1	0	0
	"Slip of tectonic plates"	1	4	24	33	56	53	13	1	0
	"Excess of heat"	0	2	4	1	1	0	0	0	1
	"After an hurricane"	0	4	3	4	3	1	0	0	0
	"I don't know"	0	17	14	13	13	2	3	0	0

Another question that has been included in the questionnaire was intended to assess, for example, how knowledgeable were the schoolchildren regarding earthquakes; in particular the following question has been asked: "Do you know why an earthquake occurs?" The respondents had to rate a five point Likert scale associated with the following: 1 = "Gods' punishment," 2 = "Slip of tectonic plates," 3 = "Excess of heat," 4 = "After an hurricane," and 5 = "I don't know." The results are shown in Table 4. It can be seen that urban schoolchildren from 9 to 12 years old (yo) agree that earthquakes are originated with a sudden slip of tectonic plates that may be considered as the correct answer to the question, with 62.8% (314/500). This was followed by those respondents that did not know the answer with 22% (110/500), 16.2% (81/500) of whom the range of ages were from 8 to 10 years old. Finally, 5% (25/500) of children considered that earthquakes are originated by God's punishment, as shown in Table 4. Similarly, the majority of the semi-rural schoolchildren responded correctly to the above question; i.e., 64.7% (185/286) of which 62.5% (179/286) were from the ages between 8 to 13 years old. On the other hand, 21.7% (62/286) did not know the answer.

Table 5. Participants' responses to the two major hazards affecting them at home and school

	Hazards	Urban schoolchildren		Semi-rural schoolchildren	
		N	%	*N*	%
School	Floods	142	28.4	77	26.7
	Hurricanes	119	23.8	109	37.7
	Earthquakes	395	79.0	209	72.3
	Landslides	66	13.2	65	22.5
	Severe weather	58	11.6	32	11.1
	Crime/Delinquency	142	28.4	64	12.1
Home	Floods	178	36.6	126	43.4
	Hurricanes	96	19.7	113	39.0
	Earthquakes	378	77.6	196	67.6
	Landslides	53	10.9	49	16.9
	Severe weather	28	7.7	34	11.7
	Crime/Delinquency	170	34.9	42	14.5

Seismic Hazard Awareness and Risk Perception

This section presents some of the results associated with schoolchildren seismic risk perception when they were asked a series of questions that addressed this issue. For example, the following: "What are the two hazards that are likely to affect you at school and at home?" The results are shown in Table 5. It can be seen that both urban and semi-rural schoolchildren seem that are well aware of the seismic risk they are exposed to at school. For example, 79% (395/487) of urban schoolchildren endorsed that earthquakes pose the biggest threat to them, followed by floods and crime with 28.4% (142/487). Similarly, 72.3% (209/297) of semi-rural schoolchildren considered earthquakes as a major threat to them. However, the second hazard that urban schoolchildren perceive that may affect them at school was hurricanes with 37.7% (109/297).

On the other hand, when the same question was asked in relation to the two hazards they face at home, both urban and semi-rural schoolchildren considered earthquakes as the major threat with 77.6% (378/487) and 67.6% (196/297), respectively. As opposed to 'crime' perceived as the second threat they face at school, both urban and semi-rural students consider that 'floods' has been considered as the second major threat they face at home with 36.6% (178/487) and 43.4% (126/297), respectively. Finally, 'severe weather' has been considered as the minor risk they are exposed at home with 7.7% (urban schoolchildren), and 11.7% (semi-rural schoolchildren). See Table 5.

In an attempt to assess the participants' perception of seismic risk events occurring in the future, the following question was put to the children: "When do you think the next earthquake will affect your community?" The results are shown in Table 6. It can be seen that 48.3% of urban schoolchildren perceive that an earthquake will occur within "Next year"; 11% ("Within the next 10 years"); 34.5% ("Within your lifetime"); only 6.3% considered that it "Never" will occur. Semi-rural schoolchildren, on the other hand, agree that an earthquake will occur "Next year" with 46.8%, 17.1% ("Within the next 10 years"), 275% ("Within your lifetime"), 9.1% considered that it "Never" will occur.

Psychological Issues and Seismic Risk Communication

Schoolchildren were asked about several questions regarding psychological issues; for example, the following "If an emergency happens, some kids and adults get upset. If you were to get upset, do you feel you would be able to help you feel less upset?" Children rated whether they perceived to be upset or scared when discussing earthquakes; the following 5-point Likert scale has been considered: 1 = "Definitely not," 2 = "Probably not," 3 = "Neutral," 4 = "Probably yes," 5 = "Definitely yes." The mean score was 3.1732 (SD = 1.393), which may be interpreted as an average urban schoolchildren response not that consistent as we have hoped for. Similar results were obtained for the case of the semi-rural schoolchildren (Mean = 2.97, SD = 1.43). This may be consistent with the frequency data shown in Table 7.

The following question has been put to the children: "Does thinking or talking about earthquakes scare or upset you?" Children rated whether they perceived to be upset or scared when discussing earthquakes; the following 3-point scale has been considered: 1 = "Not at all," 2 = "Sometimes," 3 = "Always." The mean score was 1.8865 (SD = 0.559) meaning that on average urban schoolchildren sometimes feel upset or scare when discussing earthquakes. Similar results were obtained for the case of the semi-rural schoolchildren (Mean = 1.90, SD = 0.515). This may be consistent with the frequency data shown in Table 8.

Table 6. Schoolchildren's seismic risk perception

	Hazards	"Next year"	"Within 10 years"	"Within your lifetime"	"Never"	Mean	SD	*N*
Urban schools	Floods	18.3	18.3	34.6	28.8	2.738	1.0661	459
	Hurricanes	18.4	24.3	30.3	27.0	2.658	1.0655	445
	Earthquakes	48.3	11.0	34.5	6.3	1.986	1.0394	447
	Landslides	11.9	25.3	35.4	27.4	2.782	0.9783	446
	Sever weather	16.3	24.1	39.7	20.0	2.633	0.9794	436
	Crime/ Delinquency	43.1	10.9	38.0	8.1	2.111	1.0604	432
Semi-rural schools	Floods	20.1	15.7	28.7	35.4	2.800	1.1305	254
	Hurricanes	12.5	30.1	29.7	27.7	2.726	1.0036	256
	Earthquakes	46.8	17.1	27.0	9.1	1.984	1.0522	252
	Landslides	21.4	24.0	26.6	27.9	2.611	1.1090	229
	Sever weather	24.3	21.7	32.2	21.7	2.513	1.0847	230
	Crime/ Delinquency	33.5	16.5	30.9	19.1	2.365	1.1345	230

Table 7. Participants' responses to some psychological issues regarding experiences with earthquakes

	Urban schoolchildren		Semi-rural schoolchildren	
	N	%	*N*	%
"Definitely not"	46	9.1	36	12.2
"Probably not"	169	33.3	111	37.8
"Neutral"	93	18.3	45	15.3
"Probably yes"	51	10.0	22	7.5
"Definitely yes"	149	29.3	79	26.9

The results shown in Table 8 indicates that 1 in 5 of urban schoolchildren do not get upset when discussing earthquakes; on the other hand, 1 in 10 children often feel upset on the subject. Similar trends were observed for the case of semi-rural schoolchildren, as shown in Table 8.

Table 8. Children's responses regarding the question whether they get upset when talking/discussing about earthquakes

	Urban schoolchildren		Semi-rural schoolchildren	
	n	%	n	%
"Not at all"	111	21.3	54	18.7
"Sometimes"	354	68.1	216	72.7
"Always"	54	10.4	27	9.1

Table 9. Participants' responses regarding the question whether their parents get upset when talking/discussing about earthquakes

Possible answers	Urban schoolchildren		Semi-rural schoolchildren	
	N	%	*N*	%
"No"	58	11.2	33	11.2
"I don't know"	199	38.4	121	40.7
"Yes"	261	50.4	140	47.1

In a similar vein, the following question has been put to the schoolchildren: "Does talking about seismic risk upset your parents?" Children rated whether they perceived to be upset or scared when discussing earthquakes with their parents; the following 3-point scale has been considered: 1 = "No," 2 = "I don't know," and 3 = "Yes." The results are shown in Table 9. The mean score obtained for the case of urban schoolchildren was 1.27 (SD = 0.650) meaning that on average schoolchildren endorse that their parents feel upset when discussing earthquakes with them. Similar results were obtained for the case of the semi-rural schoolchildren (Mean = 1.29, SD = 0.660). Again, the frequency data shows this pattern, see Table 9.

In addition, it can be seen that 1 out of 5 of urban schoolchildren's parents do not get upset when talking about seismic risk; on the other hand, 1 out of 10 children do not know. Similar trends were observed for the case of semi-rural schoolchildren (Table 9).

The questionnaire included questions associated with seismic risk communication at home and at school. Schoolchildren were asked to whether they talk about earthquakes with their parents or teachers. For example, the following question: "If you talked about earthquakes at school, who bring it up?" The question was rated with the following 3-point scale: 1 = "I," 2 = "Teacher," and 3 = "Both." The results are shown in Table 10. The mean score obtained from the analysis was 2.3 (SD = 0.54), which may be interpreted as an average urban schoolchildren endorse that the teacher brings the subject on at school. Similar findings were obtained for the case of the semi-rural schoolchildren (Mean = 2.39, SD = 0.511). Again, this may be consistent with the frequency data shown in Table 10.

The results also show that 1 in 20 of urban schoolchildren bring the subject up at school, in discussing earthquakes; on the other hand, 1 in 3 children endorse that both students and teachers bring on the subject. Similar trends were observed for the case of semi-rural schoolchildren (Table 10).

Similarly, the following question has been put to the children: "If you talked about earthquakes at home who bring it up?" The question was rated with the following 3-point scale: 1 = "I," 2 = "Parents," and 3 = "I and my parents." The results are shown in Table 11. The mean score obtained from the analysis was 2.2 (SD = 0.70) which may be interpreted as an average urban schoolchildren endorse that their parents brings the subject on at home. Similar findings were obtained for the case of the semi-rural schoolchildren (Mean = 2.23, SD = 0.728). Again, this may be consistent with the frequency data shown in Table 11.

The results also show that 1 in 3 of urban schoolchildren endorse that both they and their parents bring the subject on at home; on the other hand, 1 in 2 semi-rural schoolchildren endorse that both students and their parents bring the subject on at home (Table 11).

Table 10. Seismic risk communication at school

Possible answers	Urban schoolchildren		Semi-rural schoolchildren	
	N	%	*N*	%
"I"	22	4.3	3	1.0
"Teacher"	314	61.4	170	58.0
"Both"	176	34.4	120	41.0

Table 11. Seismic risk communication at home

	Urban schoolchildren		Semi-rural schoolchildren	
	n	%	n	%
"I"	83	16.4	50	17.5
"My parents"	239	47.1	119	41.6
"I and my parents"	184	36.3	117	40.9

Table 12. Preparedness on seismic risk

Possible answers	Urban schoolchildren		Semi-rural schoolchildren	
	N	%	*N*	%
"No"	27	5.2	23	8.0
"Yes"	488	94.6	264	92.0
Total	515	100.0	487	100.0

Preparedness

A number of questions were put to children regarding the issue associated with knowledge on preparedness. For example: "Do you know what to do during an earthquake when you are in the classroom?" which was put to the children on a 2-point scale ("Yes" = 1 and "No" = 0). The results are shown in Table 12. It can be seen that the vast majority of the urban schoolchildren consider that they know what to do during an earthquake in the classroom; i.e., 94% (488/515) with a Mean = 0.9535 and a SD = 0.26026. (It should be noted that in the school where the data was collected two children shared a desk and 'to know what to do' implies to cover under it in case of a quake). Similar results were obtained with semi-rural schoolchildren; i.e., 92% (264/287) with a Mean = 0.9199 and a SD = 0.27198.

DISCUSSION

A questionnaire (i.e., 'instrument') intended to assess the degree of knowledge, risk perception, psychological and communication issues regarding seismic risk has been designed. The 'instrument' consists of a total of 60 items and have been derived from a number of studies being published in the literature (McClure, Wills, Johnston & Recker, 2011; Lindell, 1994; Beck, etal., 2012; Marincioni, Appiotti, Ferretti, Antinori, Melonaro, Pusceddu & Oreficini-Rosi, 2012; Pan, 2012; Ronan & Johnston, 1996; Ronan, et al., 2001; Zhang et al. 2012; and those reported by Santos-Reyes, et al., 2014a,b). However, as stated in the introduction section, there are not that many studies being conducted on schoolchildren's seismic risk perception with the exception of a study reported by Ronan et al. (2001). In particular, questions associated with psychological and communication issues have been adopted in the present study from the study reported by Ronan et al. (2001). The paper has presented the results of some of the items included in the questionnaire but not all of them. In addition, the results are presented from a descriptive statistics perspective. The inferential analysis has been reported elsewhere (Santos-Reyes et al., 2016).

Overall, the results presented have shown that a relatively high of percentage of urban schoolchildren are well aware that they live in a region prone to earthquakes (Mean = 0.63; SD = 0.48); semi-rural schoolchildren, on the other hand, rated low on this issue (Mean = 0.44; SD = 0.49). The results were quite surprising given the fact that there have been a number of earthquakes occurring in the region in recent years; for example, from December 2011 to May 2014 there have been a total of 10 quakes of considerable magnitudes; i.e., ranging from 5.9 to 7.5 on the Richter scale (Santos-Reyes et al., 2014a,b). In general, it may be argued that there is a lack of seismic educational campaign in the region under study. In fact, this is the first study being conducted in the region. In addition, this trend (i.e., lack of awareness of the seismic region where the participants live) is consistent with the results associated with the degree of knowledge on earthquakes reported, for example, in Table 3. Effectively, an educational campaign to further educate both urban and semi-rural schoolchildren is urgently needed in this particular region of the country.

The two hazards associated with earthquakes and crime have been perceived to represent the major threats at school and at home for both urban and semi-rural schoolchildren (Table 5). Crime may be explained given the fact that recently the organised crime committed by the drug cartels is very much in the news every day (Santos-Reyes et al. 2014a,b). On the other hand, severe weather has been rated as an event unlikely to occur at school (i.e., urban school children: 11.6%; semi-rural school children: 11.1%) and at home (urban schoolchildren: 7.7%; semi-rural schoolchildren: 11.7%). These results show the trend found by Slovic (1987) and Ronan et al., (2001) that "children were able to identify higher frequency events at a higher frequency and lower frequency events at a relatively lower frequency" (Ronan, et al., 2001). On the other hand, "adult samples have been found to identify high impact/low frequency events at a higher frequency than low impact/high frequency events" (Ronan, et al., 2001).

Children rated whether they perceived to be upset or scared when discussing earthquakes; the mean score was 3.1732 (SD = 1.393) which may be interpreted as an average urban school children response not that consistent as we have hoped for. Similar results were obtained for the case of the semi-rural schoolchildren (Mean = 2.97, SD = 1.43). This may be explained given the need to get involved in seismic risk educational programmes.

Communication between children and parents, school teachers and vice versa plays a crucial role in learning more, inter alia, of what to do well before, during and after earthquakes. The results show that effectively seismic risk communication takes place at school (where the teacher brings the subject on, Mean = 2.3; SD = 0.54, for example, the case of urban schoolchildren) and at home (where the parents bring the topic on with a Mean = 2.2 and SD = 0.7 and for the case of urban schoolchildren). It should be pointed out that the findings are encouraging, given the fact that children can educate further their parents about earthquakes. That is, most of the semi-rural families are not well educated and this may help to promote the culture of prevention regarding seismic risk (Santos-Reyes et al., 2014a,b; Marincioni et al., 2012; Beck et al., 2012; Ronan et al., 2001).

Last but not least, it has been argued elsewhere (Santos-Reyes & Beard 2011) that injury or fatality due to disaster risk, including seismic risk, represents a failure of the 'system' which, ultimately, encompasses all of society. Because of this it is vital to have an approach which considers things as a whole in the widest sense and such an approach may be called 'systemic.' A model has been put foreword from this perspective, and the results presented here (and similar studies) should be seen as a part of such an approach. That is, in order to be able to build a resilience community to seismic risk it is necessary to have an effective disaster management system.

CONCLUSION

The main results are the following:

a) Overall, urban school children demonstrated a generalized awareness of the seismic area where they live (Mean = 0.63; SD = 0.48); however, semi-rural schoolchildren was not the case (Mean = 0.44; SD = 0.49); educational campaigns may be needed to further enhance seismic awareness of both urban and semi-rural areas of the region.
b) The two hazards associated with earthquakes and crime have been perceived to represent the two major threats to them at school and at home, and for both urban and semi-rural schoolchildren.

c) Schoolchildren emotions are not consistent when asked whether they would be able to control themselves during an emergency.
d) Seismic risk communication takes place at school and at home; this may be regarded as encouraging in order to further educate children on seismic risk; however, more is needed to better educate them on important aspects regarding what to do before, during and after an earthquake and this may well prove to be an effective way to share their knowledge with their parents (most of them are not well educated) and contribute to built a resilient community to earthquakes.
e) The preliminary findings of the study are consistent with has been found in similar studies; i.e., "children were able to identify higher frequency events at a higher frequency and lower frequency events at a relatively lower frequency" (Ronan et al., 2001).

ACKNOWLEDGMENTS

This project was funded under the following grants: CONACYT, No-248219 and SIP-IPN: No-20161496.

REFERENCES

Beck, E., André-Poyaud, I., Paule-Annick, D., Chardonnel, S., and Lutoff, C. (2012). Risk perception and social vulnerability to earthquakes in Grenoble (French Alps), *Journal of Risk Research*, 15(10), 1245-1260.

Blaikie, P., Cannon, T., Davis, I., and Wisner, B. (1994). At risk: natural hazards, people's vulnerability, and disasters. London, Routledge.

Contreras, D., Blaschke, T., Kienberger, S., and Zeil, P., 2014. Myths and realities about the recovery of L'Aquila after the earthquake. *International Journal of Disaster Risk Reduction*, 8, 125-142.

Drabeck, T. (1986). *Human System Responses to Disasters: An Inventory of Sociological Findings.* New York, Springer-Verlag.

Higgins, E.T., and Bargh, J. A. (1987). Social cognition and social perception. *Annual Review of Psychology*, 38, 369-425.

Kahnemen and Tversky. (1973). On the psychology of prediction. *Psychological Review*, 80(4), 237-251.

Lermo J et al. (1991). *Microzonificación Sísmica y Análisis de amplificación relativa en la ciudad de Oaxaca*. Mexico, CENAPRED. [Seismic micro-zoning and analysis relative amplification in the city of Oaxaca. CENAPRED, Mexico].

Lindell, M. K. (1994). Perceived Characteristics of Environmental Hazards. *International Journal of Mass Emergencies and Disasters*, 12(3), 303-326.

Marincioni, F., Appiotti, F., Ferretti, M., Antinori, C., Melonaro, P., Pusceddu, A., and Oreficini-Rosi, R. 2012. Perception and Communication of Seismic Risk: The 6 April 2009 L'Aquila Earthquake Case Study. *Earthquake Spectra*, 28(1), 159-183.

McClure, J., Wills, C., Johnston, D., and Recker, C. (2011). How the 2010 Canterbury (Darfield) earthquake affected earthquake risk perception: Comparing citizens inside and outside the earthquake region. *Australasian Journal of Disaster and Trauma Studies*, 2011-2, 1-10.

Nuñez-Cornú, F., and Ponce, L. (1998). Zonas sísmicas de Oaxaca, Mexico: Sísmos máximos y tiempos de recurrencia para el periodo 1542-1988. *Geofísica Internacional*, 28(4), 587-641. [Seismic zones of Oaxaca, Mexico: Seismic maximum and recurrence times for the period 1542-1988].

Pan, A. (2012). A Study on Residents' Risk Perception in Abrupt Geological Hazard. *Journal of Risk Analysis and Crisis Response*, 2(1), 44-55.

Paton, D., Anderson, E., Becker, J., Petersen, J. (2015). Developing a comprenhensive model of hazard preparedness: Lessons from the Christchurch earthquake. *International Journal of Disaster Risk Reduction*, vol. 14(1): 37-45.

Raby, A., Macabuag, J., Pomonis, A., Wilkinson, S., and Rossetto, T., 2015. Implications of the 2011 Great East Japan Tsunami on sea defence design. *International Journal of Disaster Risk Reduction*, 14(4), 332-346.

Ronan, K. R. and Johnston, D. M. (1996). Volcanic eruptions and children: Assessment and school based intervention. Paper presented at the Pan Pacific Hazards '96 Conference, Vancouver.

Ronan, K.R., Johnston, D.M., Daly, M., and Fairley, R. (2001). School children's risk perception and preparedness: a hazard education survey. *The Australasian Journal of Disaster and Trauma Studies*, 1, 1-35.

Santos-Reyes J. and Beard A.N. (2011). Applying the SDMS model to the analysis of the Tabasco flood disaster in Mexico, *Human and Ecological Risk Assessment; An International Journal*, 17(3), 646-677.

Santos-Reyes, J., Gouzeva, T, and Santos-Reyes, G. (2014a). Earthquake risk perception and communication: A review of empirical research. *Disaster Advances* 7(7):77-87.

Santos-Reyes, J., Gouzeva, T., and Santos-Reyes, G. (2014b). Earthquake risk perception and Mexico City's public safety. *Procedia Engineering*, 84, 662-671.

Santos-Reyes, J., Santos-Reyes, G., Gouzeva, T., and Velazquez-Martinez, D. (2016). Schoolchildren's earthquake knowledge, preparedness and risk perception of a seismic prone region of Mexico. *Human and Ecological Risk Assessment: An International Journal*. In press.

Sims, A. (2016). *The quake struck in the early hours of the morning when most residents were asleep*. The Independent, Retrieved from http://www.independent.co.uk/news/world/europe/italyearthquaketodaydeathtolllatest newsamatriceaccumolimagnitudedevastationa7208046.html.

Slovic, P. (1987). Perception of risk. *Science*, 236, 280-285.

Slovic, P. Fishhoff, B., and Lichtenstein, S. (1981). *Perceived risk: psychological factors and social implications*. Proceedings of the Royal Society of London A376, 17-34.

Smith, R.M., Dyer, G.S., Antonageli, K., Arredondo, N., Bedilion, H., Dalal, A., Deveny, G.M., Joseph, G., Lauria, D., Lockhart, S.H., Lucien, S., Marsh, S., Rogers, S.O., Salzarulo, H., Shah, S., Toussaint, R.J., and Wagoner, J. (2012). Disaster triage after the Haitian earthquake. *Injury*, 43 (11), 1811-1815.

Sparf, J., and Öhman, S. (2014). On Risk and Disability – Investigating the Influence of Disability and Social Capital on the Perception and Digital Communication of risk. *Journal of Risk Analysis and Crisis Response*, 4 (1), 20-33.

In: Earthquakes
Editor: Wayne Coleman

ISBN: 978-1-53610-342-7

Chapter 7

Anomalous Decrease in Groundwater Radon Before the Taiwan Large Earthquakes

T. Kuo*

Department of Mineral and Petroleum Engineering,
National Cheng Kung University, Tainan, Taiwan

Abstract

In southwestern Taiwan, radon anomalous declines prior to the 2010 Mw 6.3 Jiasian and 2016 Mw 6.4 Meinong earthquakes were recorded at the Paihe spring. The above-mentioned earthquakes are the two strongest since 1964 that have occurred near the area southeast of Chiayi. The Paihe spring (P1) is located 46 km and 45 km, respectively, from the epicenters of the 2010 Mw 6.3 Jiasian and 2016 Mw 6.4 Meinong earthquakes. Specifically, the concentration of groundwater radon decreased from background levels of 144 ± 7 and 137 ± 8 pCi/L to minima of 104 ± 8 and 97 ± 9 pCi/L prior to the 2010 Mw 6.3 Jiasian and 2016 Mw 6.4 Meinong earthquakes, respectively. The Paihe spring is from an unconfined limestone aquifer surrounded by ductile shale and sandy shale. The v-shaped pattern of radon behavior in the Paihe spring is quite similar to those radon decreases that were measured in the confined andesite aquifer at Antung in eastern Taiwan.

* Corresponding Author address Email: mctkuobe@mail.ncku.edu.tw.

Recurrent anomalous declines in the concentration of groundwater radon were observed at well D1 in the Antung hot spring in eastern Taiwan prior to the five major earthquakes –2003 Mw 6.8 Chengkung, 2006 Mw 6.1 and Mw 5.9 Taitung, 2008 Mw 5.4 Antung, and 2011 Mw 5.0 Chimei. Well D1 is located 24 km, 52 km, 47 km, 13 km, and 32 km, respectively, from the epicenters of the 2003 Mw 6.8, 2006 Mw 6.1 and Mw 5.9, 2008 Mw 5.4, and 2011 Mw 5.0 earthquakes. The 2003 Mw 6.8 Chengkung was the strongest earthquake near the Chengkung area in eastern Taiwan since 1951. Specifically, radon decreased from a background level of 787 ± 42 pCi/L to a minimum of 326 ± 9 pCi/L prior to the 2003 Mw 6.8 Chengkung earthquake. In a brittle aquifer under un-drained conditions (Antung and Paihe), the dilation of brittle rock mass occurred at a rate faster than the recharge of pore water and gas saturation developed in newly created cracks preceding the earthquake. Radon partitioning into the gas phase may explain the anomalous decrease of groundwater radon prior to the 2003 Chengkung earthquake.

Both observations at Paihe and Antung suggest that in-situ radon volatilization offers an attractive mechanism for a premonitory decrease in groundwater radon. "A low-porosity brittle aquifer with un-drained conditions" is a suitable geological site to consistently detect precursory declines in groundwater radon prior to local large earthquakes. A long-term monitoring of precursory declines in groundwater radon at a suitable geological site can be a useful means for forecasting local disastrous earthquakes. At a suitable geological site, it is of a great practical value to develop the correlations among the scale of radon decline, precursory time, and earthquake magnitude for a given seismogenic fault.

Keywords: radon anomaly, groundwater, earthquake, magnitude, precursory time

INTRODUCTION

Measurement of radon concentration (^{222}Rn) in groundwater has been frequently used in earthquake prediction (Igarashi et al., 1995; Noguchi and Wakita, 1977; Wakita et al., 1980). According to a worldwide survey (Hauksson, 1981), more than 80 % of radon anomalies measured in groundwater prior to earthquakes show increased radon concentration, whereas a few anomalies were characterized by decreased radon concentration. For example, anomalous decreases in radon concentration were observed in groundwater prior to the 1978 Izu-Oshima-Kinkai earthquake of magnitude *M* 7.0 in Japan (Wakita et al., 1980) and prior to the 2003

Chengkung earthquake of magnitude M_w 6.8 in Taiwan (Kuo et al., 2006a). The distance from the epicenter to the radon monitoring station (well D1) for the 2003 Chengkung earthquake was about 20 km. Similarly, the distance from the epicenter to the radon monitoring station (well SKE-1) for the 1978 Izu-Oshima-Kinkai earthquake was about 25 km. Sixty-five days before the 2003 Chengkung earthquake, the radon concentration of groundwater started to decrease. Similarly, about eighty days before the 1978 Izu-Oshima-Kinkai earthquake, the radon concentration started to decrease. Our research effort focuses on the anomalous declines in groundwater radon and their mechanisms precursory to large earthquakes in Taiwan.

We have studied recurrent radon declines in eastern Taiwan for nearly 14 years. In July 2003, we initiated an observation of groundwater radon at the Antung hot spring (D1). Well D1, located in eastern Taiwan, is about 3 km southeast of the Chihshang fault. The fault ruptured during two 1951 earthquakes of magnitudes M 6.2 and M 7.0 (Hsu 1962). The Chihshang fault is part of the eastern boundary of the present-day plate suture between the Eurasia and the Philippine plates. Since July 2003, recurrent groundwater radon anomalous declines were observed at Well D1 in the Antung hot spring in eastern Taiwan prior to the five earthquakes of magnitude M_w 6.8, M_w 6.1, M_w 5.9, M_w 5.4, and M_w 5.0 that occurred on December 10, 2003, April 1, 2006, April 15, 2006, February 17, 2008, and July 12, 2011, respectively. We consider the M_w 5.9 earthquake that occurred on April 15, 2006 triggered by stress transfer in response to the 2006 M_w 6.1 Taitung earthquake. Well D1 is located 24 km, 52 km, 13 km, and 32 km, respectively, from the epicenters of the 2003 M_w 6.8, 2006 M_w 6.1, 2008 M_w 5.4, and 2011 M_w 5.0 earthquakes. Prior to the above major earthquakes, radon concentration decreased from background levels of 787 ± 42, 762 ± 57, 700 ± 57, and 752 ± 24 pCi/L to minima of 326 ± 9, 371 ± 9, 480 ± 43 pCi/L, and 447 ± 18 pCi/L, respectively. The observed precursory minimum in radon concentration decrease as the local earthquake magnitude associated with the Chihshang fault increases (Kuo et al., 2010a, 2011). The precursor time (time interval between the moment when the trend of the data starts to decline and the time of occurrence of the earthquake) increases as the local earthquake magnitude increases (Kuo, 2014). Monitoring precursory decline in groundwater radon at well D1 in the Antung hot spring has been successfully demonstrated to be a useful means of forecasting the magnitude and precursor time of local disastrous earthquakes in eastern Taiwan (Kuo, 2014).

Regarding a physical basis that explains the characteristic radon decline, an anomalous decrease in radon concentration in groundwater prior to the

2003 Chengkung earthquake, Kuo et al., (2006b) presented an in-situ volatilization mechanism based on radon phase behavior and the geological conditions of the Antung hot spring. The Antung hot spring is a low-porosity fractured isolated aquifer situated in an andesitic block and surrounded by a ductile mudstone of the Lichi mélange (Chen and Wang 1996). Under these geological conditions, the development of new cracks in the aquifer rock could occur at a rate faster than the recharge of pore water. In an aquifer with such un-drained conditions, gas saturation and two phases (vapor and liquid) could develop in the rock cracks (Brace et al., 1966; Nur 1972; Scholz et al., 1973; Scholz 1978). Meanwhile, the radon in groundwater volatilizes and partitions into the gas phase and the concentration of radon in groundwater decreases. The mechanism of in-situ radon volatilization was substantiated by the simultaneous anomalous declines in groundwater-dissolved radon and methane precursory to the 2008 M_w 5.4 Antung earthquake (Kuo et al., 2010b) and also by the simultaneous concentration declines in groundwater radon, methane, and ethane precursory to the 2011 M_w 5.0 Chimei earthquake (Kuo et al., 2013).

Based on the mechanism of in-situ volatilization and the local geological conditions near the Antung hot spring of eastern Taiwan, we proposed a somewhat similar monitoring site (a low-porosity brittle aquifer with un-drained conditions) in southwestern Taiwan to consistently catch precursory declines in groundwater radon. The objective of this chapter is to demonstrate that radon decreases prior to large earthquakes in eastern Taiwan that were observed in an aquifer in eastern Taiwan can also be observed in a spring under similar geological conditions in southwestern Taiwan. The Paihe spring is an unconfined limestone aquifer surrounded by mudstone of the Yunshuichi Formation (Yen and Chang 1949), whereas the Antung hot spring is a confined andesitic aquifer in the Lichi mélange. In November 2009, we initiated an observation of groundwater radon at the Paihe spring (P1). The 2010 M_w 6.3 Jiasian and 2016 M_w 6.4 Meinong earthquakes are the strongest since 1964 near the area southeast of Chiayi (Figure 1). The purpose of this chapter is to present the radon anomalous declines recorded at the Paihe spring (P1) in southwestern Taiwan and at the Antung hot spring (Well D1) in eastern Taiwan prior to the 2010 M_w 6.3 Jiasian, 2016 M_w 6.4 Meinong, and 2003 M_w 6.8 Chengkung earthquakes.

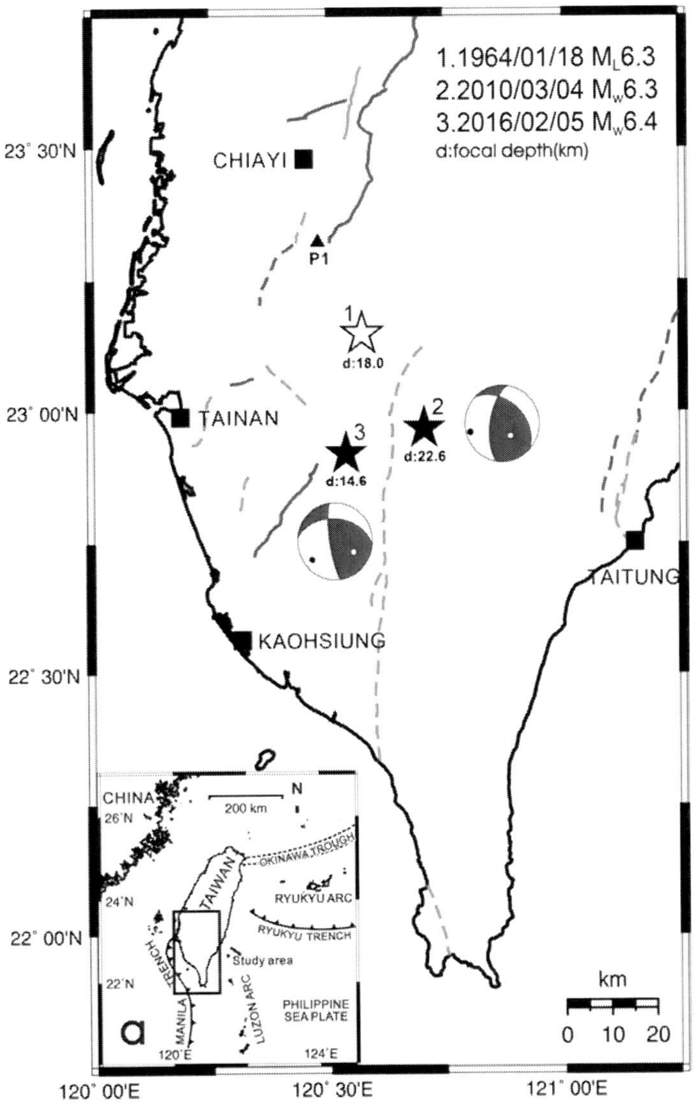

Figure 1. Map of the epicenters of the earthquakes that occurred on January 8, 1964, March 4, 2010, and February 5, 2016 in southwestern Taiwan. (a) Map of Taiwan. (b) Study area (open star: 1964 mainshock, filled stars: 2010 Jiasian and 2016 Meinong mainshocks, filled triangle: Paihe spring P1).

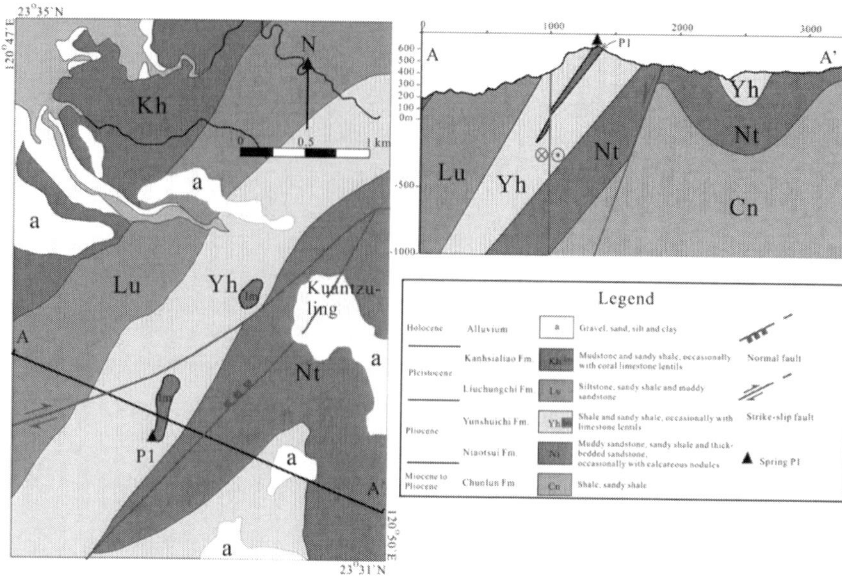

Figure 2. Geological map and cross section near the Paihe spring (P1).

GEOLOGY OF THE PAIHE SPRING AND THE ANTUNG HOT SPRING

The Paihe spring is from a brittle limestone aquifer surrounded by ductile shale and sandy shale of the Yunshuichi Formation (Figure 2). The limestone forms an isolated, flat-topped peak called Chentou Mountain, which is about 646 m above sea level and looks like a pillow from a distant view (Yen and Chang 1949). The limestone is essentially built of reef corals, generally massive. The limestone lentils strike N 30° ~ 50° E, dipping 50° ~ 60° to NW. The limestone extends about 900 m with an average thickness of 40 m. Precipitation infiltrates into the limestone lentils. The aquifer recharge only occurs during the rainy season. The dry season usually starts from end of November to end of April next year. During the dry season, the unconfined limestone aquifer at Chentou Mountain can be regarded in un-drained conditions. Environmental factors particularly rainfall affect the radon concentration of groundwater samples taken from unconfined aquifers. There is no heavy rainfall in the dry season to affect the radon concentration of groundwater samples taken from the Paihe spring (P1).

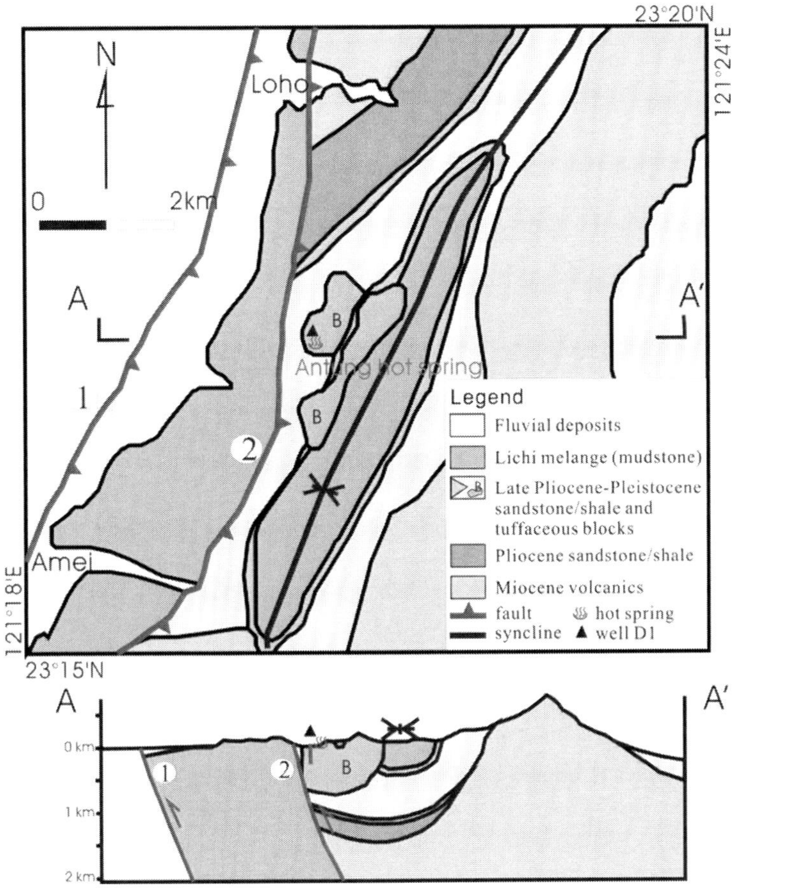

Figure 3. Geological map and cross section near the radon-monitoring well D1 in the area of Antung hot spring. (B: tuffaceous andesitic blocks; D1: radon-monitoring well; ①: Chihshang, or, Longitudinal Valley Fault, ②: Yongfeng Fault).

The Antung hot spring is situated in an exotic brittle tuffaceous-sandstone block surrounded by a ductile mudstone of the Paliwan Formation and the Lichi mélange (Chen and Wang 1996). The exotic block extends about 1 or 2 km (Figure 3). The hot spring is formed nearby an eastward-dipping, high-angle reverse fault zone which contacts between the Lichi mélange and the Paliwan Formation. Some hot springs and mud volcanoes are scattered along the fault zone, indicating a Quaternary active fault. Well D1 is not artesian and the aquifer near Well D1 can be regarded in un-drained conditions. The Antung hot spring is a confined aquifer. Environmental factors such as atmospheric temperature, barometric pressure, and rainfall do not affect the radon concentration of groundwater samples taken from Well D1.

In a small brittle aquifer under un-drained conditions (Antung Well D1 or Paihe spring P1), we hypothesize that gas saturation could develop in the rock cracks when the regional stress increased. Then, the radon in the groundwater volatilized into the gas phase, thereby causing a decrease in groundwater radon concentration (Kuo et al., 2006b).

RADON-MONITORING METHODS

Discrete water samples were collected at the Paihe spring (P1) and Antung Well D1 about twice per week for analysis of radon content. Water samples were collected in a 40 mL glass vial with a TEFLON-lined cap. The 40 mL glass vial was filled and capped underwater near the spring outlet. After collecting a sample, the sample vial was inverted to check for air bubbles. If any bubbles were present in the vial, the sample water was discarded and sampling was repeated. The date and time of sample collection were recorded. The samples were stored and transported in a cooler. Radioactivity counts were completed within 4 days for each sample.

The liquid scintillation method was adopted to determine the activity concentration of radon in groundwater (Noguchi, 1964). Radon was partitioned selectively into a white mineral-oil based scintillation cocktail (Perkin Elmer) immiscible with the water samples, and then assayed with a liquid scintillation counter (LSC). The results were corrected for the amount of radon decay between sampling and assay.

A calibration factor for the LSC measurements of 7.0 ± 0.3 cpm/pCi was calculated using an aqueous ^{226}Ra calibration solution, which is in secular equilibrium with the ^{222}Rn progeny. For a count time of 50 min and background level of 4.8 ± 0.6 cpm, a detection limit below 14 pCi/L was achieved using the sample volume of 15 ml.

RECURRENT ANOMALOUS DECREASES IN GROUNDWATER RADON BEFORE THE TAIWAN LARGE EARTHQUAKES

The monitoring of groundwater radon was initiated at the Paihe spring (P1) in southwestern Taiwan in November 2009. Radon anomalous declines prior to the 2010 M_w 6.3 Jiasian and 2016 M_w 6.4 Meinong earthquakes were recorded at the Paihe spring (P1). The two above-mentioned earthquakes are the strongest since 1964 that have occurred near the area southeast of Chiayi. The Paihe spring (P1) is located 46 km and 45 km, respectively, from the epicenters of the 2010 M_w 6.3 Jiasian and 2016 M_w 6.4 Meinong earthquakes. Specifically, the concentration of groundwater radon decreased from background levels of 144 ± 7 and 137 ± 8 pCi/L to minima of 104 ± 8 and 97 ± 9 pCi/L prior to the 2010 M_w 6.3 Jiasian and 2016 M_w 6.4 Meinong earthquakes, respectively (Figure 4 and Figure 5). Radon-concentration errors are ±1 standard deviation after simple averaging of triplicates. Environmental records such as atmospheric temperature, barometric pressure, and rainfall were examined to make sure that the above radon anomalies were not attributed to these environmental factors. Rainfall particularly affects the radon concentration of groundwater samples taken from unconfined aquifers. There was no heavy rainfall responsible for the above radon anomalies. The precursor time for radon anomalous decline is defined as the time interval between the moment when the trend of the data starts to decline and the time of occurrence of the earthquake. Figure 4 and Figure 5 show that the precursor times for radon anomalous decline were 80 and 54 days prior to the 2010 M_w 6.3 Jiasian and 2016 M_w 6.4 Meinong earthquakes, respectively.

The anomalous decreases in Figure 4 and Figure 5 observed at the Paihe spring (P1) follow the same v-shaped progression as those observed at Antung. The sequence of events for radon anomalies prior to the 2010 M_w 6.3 Jiasian and 2016 M_w 6.4 Meinong earthquakes were characterized with three stages. During Stage 1, the radon concentration in ground water was fairly stable and there was an accumulation of tectonic strain, which produced a slow, steady increase of effective stress. During Stage 2, dilation of brittle rock masses occurred and gas saturation developed in rock cracks. The radon in ground water volatilized into the gas phase. The radon concentration of ground water started to decrease and reached a minimum value. Stage 3 started at the point of minimum radon concentration; the water saturation in rock cracks and pores began to increase again. During Stage 3, the radon concentration in ground water increased and recovered to the previous background level before the earthquake. Both Antung and Paihe are small fractured aquifers surrounded by ductile mudstone. Under suitable geological conditions such as Antung and Paihe, groundwater radon can be a sensitive tracer for strain changes in the crust associated with earthquake occurrences. A long-term monitoring of precursory declines in groundwater radon at a suitable geological site can be a useful means for forecasting local disastrous earthquakes.

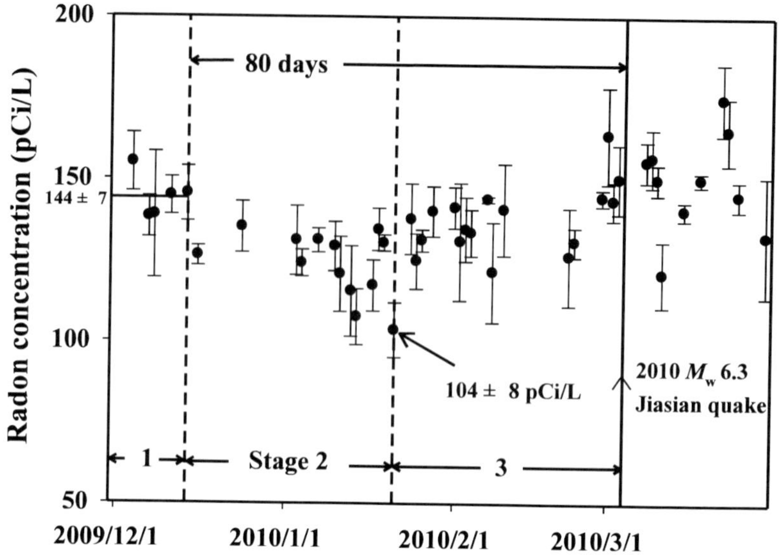

Figure 4. Observed radon anomalies at the Paihe spring (P1) prior to 2010 Jiasian earthquake. Stage 1 is buildup of elastic strain. Stage 2 is development of cracks. Stage 3 is influx of groundwater.

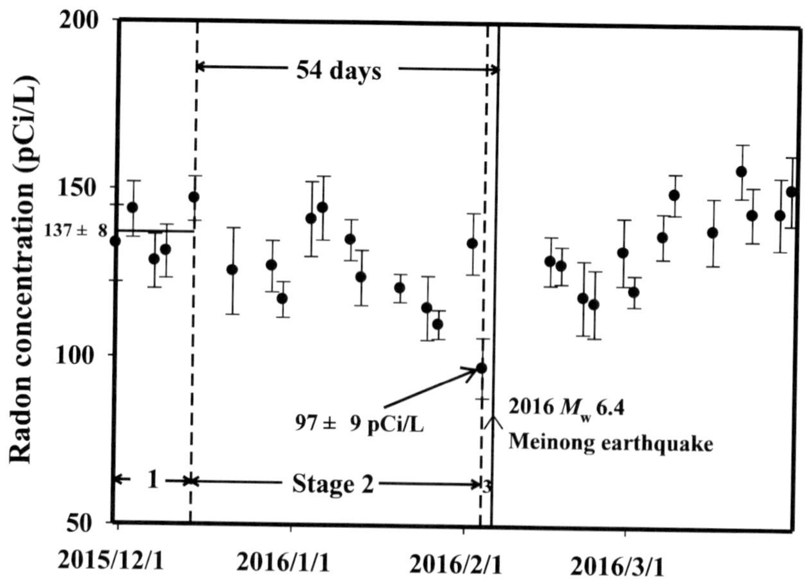

Figure 5. Observed radon anomalies at the Paihe spring (P1) prior to 2016 Meinong earthquake. Stage 1 is buildup of elastic strain. Stage 2 is development of cracks. Stage 3 is influx of groundwater.

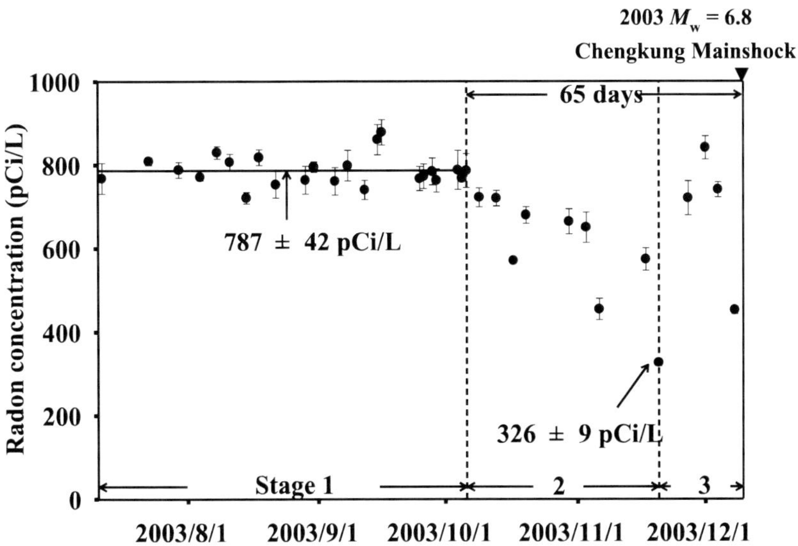

Figure 6. Observed radon anomalies at the monitoring well (D1) in the Antung hot spring prior to 2003 Chengkung earthquake. Stage 1 is buildup of elastic strain. Stage 2 is development of cracks. Stage 3 is influx of groundwater (adapted from Kuo et al. 2006b).

The 2003 M_w 6.8 Chengkung earthquake was the strongest since 1964 that has occurred near the Chengkung area in eastern Taiwan. The 2010 M_w 6.3 Jiasian and 2016 M_w 6.4 Meinong earthquakes were the strongest since 1964 that have occurred near the area southeast of Chiayi in western Taiwan. Some of the characteristics of the radon concentrations observed for the 2003 Chengkung earthquake can be compared to those of the 2010 Jiasian and 2016 Meinong earthquakes. For example, the Antung well D1 is located 24 km from the epicenter of the 2003 Chengkung earthquake; the Paihe spring (P1) is located 46 km and 45 km, respectively, from the epicenters of the 2010 Jiasian and 2016 Meinong earthquakes. The observed radon anomaly precursory to the 2003 Chengkung earthquake also behaved similarly to those observed precursory to the 2010 Jiasian and 2016 Meinong earthquakes. Figure 6 shows the anomalous decline observed at the Antung well D1 prior to the 2003 M_w 6.8 Chengkung earthquake. The concentration of groundwater radon decreased from a background level of 787 ± 42 pCi/L to a minimum of 326 ± 9 pCi/L. Sixty-five days before the 2003 Chengkung earthquake, the radon concentration of ground water started to decrease (Figure 6). Similarly, eighty and fifty-four days before the 2010 Jiasian and 2016 Meinong earthquakes, the radon concentration started to decrease, respectively (Figure 4 and Figure 5). The v-shaped pattern of radon behavior observed at Paihe spring P1 is quite similar to those radon decreases that were measured at Antung Well D1.

The v-shaped radon pattern recognized in the above radon anomalies is valuable for detecting the aseismic strain precursory to local strong earthquakes ($M_w > 6.0$). We have been monitoring recurrent radon declines at Antung Well D1 and Paihe spring P1 since July 2003 and November 2009, respectively. At a suitable geological site, a long-term monitoring of precursory declines in groundwater radon is a useful means for forecasting local disastrous earthquakes. It is of a great practical value to develop the characteristic correlations among the scale of radon decline, precursory time, and earthquake magnitude for a given site and for a given seismogenic fault.

CONCLUSION

The significance of our research can be summarized as follows.

1) The mechanism of in-situ radon volatilization has been corroborated by the radon anomalous declines in groundwater consistently recorded prior to local large earthquakes near the Paihe spring and the Antung hot spring in Taiwan. In-situ radon volatilization is an attractive mechanism for a premonitory decrease in groundwater radon.
2) "A low-porosity brittle aquifer with un-drained conditions" is a suitable geological site to consistently detect precursory declines in groundwater radon prior to local large earthquakes.
3) The Paihe spring is an unconfined limestone aquifer, whereas the Antung hot spring is a confined andesitic aquifer. Environmental factors such as atmospheric temperature, barometric pressure, and rainfall do not affect the radon concentration of groundwater samples taken from confined aquifers. On the other hand, rainfall affects the radon concentration of groundwater samples taken from unconfined aquifers. In the dry season, there is no heavy rainfall to affect the radon concentration of groundwater samples taken from unconfined aquifers.
4) At a suitable geological site, it is very valuable to develop the correlations among the scale of radon decline, precursory time, and earthquake magnitude for a given seismogenic fault. A long-term

monitoring of precursory declines in groundwater radon at a suitable geological site is a useful means for forecasting local disastrous earthquakes.

ACKNOWLEDGMENTS

Supports by the National Science Council, Central Geological Surveys, Industrial Technology Research Institute (L550001060, N550003318), Radiation Monitoring Center, and Institute Earth Sciences of Academia Sinica of Taiwan are appreciated. The author is grateful to Mr. C. Lin of the Antung hot spring and Mr. B. L. Wu of the Paihe spring for their field assistance. The author is also grateful to Dr. C. Lewis for editing for English usage.

REFERENCES

Brace, W. F., Paulding, B. W., Jr., Scholz, C. H., (1966). Dilatancy in the fracture of crystalline rocks. *J. Geophys. Res.* 71, 3939-3953.

Chen, W. S., Wang, Y., (1996). Geology of the Coastal Range, eastern Taiwan. *Geology of Taiwan* 7. Central Geological Survey, Taiwan.

Hauksson, E., (1981). Radon content of groundwater as an earthquake precursor: Evaluation of worldwide data and physical basis. *J. Geophys. Res.* 86, 9397-9410.

Hsu, T. L., (1962). Recent faulting in the Longitudinal Valley of eastern Taiwan. *Mem. Geol. Soc. China* 1, 95-102.

Igarashi, G., Saeki, S., Takahata, N., Sumikawa, K., Tasaka, S., Sasaki, Y., Takahashi, M., Sano, Y., (1995). Ground-water radon anomaly before the Kobe earthquake in Japan. *Science* 269, 60-61.

Kuo, T., Fan, K., Kuochen, H., Han, Y., Chu, H., Lee, Y., (2006[a]). Anomalous decrease in groundwater radon before the Taiwan M6.8 Chengkung earthquake. *J. Environ. Radioact.* 88, 101-106.

Kuo, T., Fan, K., Kuochen, H., Chen, W., (2006b). A Mechanism for Anomalous Decline in Radon Precursory to an Earthquake. *Groundwater* 44, 642-647.

Kuo, T., Lin, C., Chang, G.., Fan, K., Cheng, W., Lewis, C., (2010a). Estimation of aseimic crustal-strain using radon precursors of the 2003 *M*

6.8, 2006 *M* 6.1, and 2008 *M* 5.0 earthquakes in eastern Taiwan. *Nat. Hazards* 53, 219-228.

Kuo, T., Cheng, W., Lin, C., Fan, K., Chang, G.., Yang, T., (2010b). Simultaneous declines in radon and methane precursory to 2008 M_w 5.0 Antung earthquake: Corroboration of in-situ volatilization. *Nat. Hazards* 54, 367-372.

Kuo, T., Lin, C., Su, C., Liu, C., Lin, C. H., Chang, C., Chiang, C., (2011). Correlating recurrent radon precursors with local earthquake magnitude and crust strain near the Chihshang fault of eastern Taiwan. *Nat. Hazards* 59, 861-869.

Kuo, T., Liu, C., Su, C., Chang, C., Chen, W., Chen, Y., Lin, C., Kuochen, H., Hsu, Y., Lin, Y., Huang, Y., Lin, H., (2013). Concurrent concentration declines in groundwater-dissolved radon, methane and ethane precursory to 2011 M_w 5.0 Chimei earthquake. *Radiat. Meas.* 58, 121-127.

Kuo. T., (2014). Correlating precursory declines in groundwater radon with earthquake magnitude. *Groundwater* 52, 217-224.

Noguchi, M., (1964). New method of radon activity measurement with liquid scintillator. *Radioisotopes* 13, 362-367.

Noguchi, M., Wakita, H., (1977). A method for continuous measurement of radon in groundwater for earthquake prediction. *J. Geophys. Res.* 82, 1353-1357.

Nur, A., (1972). Dilatancy, pore fluids, and premonitory variations of ts/tp traval times. *Bull. Seismol. Soc. Am.* 62, 1217-1222.

Scholz, C. H., Sykes, L. R., Aggarwal, Y. P., (1973). Earthquake prediction: A physical basis. *Science* 181, 803-810.

Scholz, C. H., (1978). Velocity anomalies in dilatant rock. *Science* 201, 441-442.

Wakita, H., Nakamura, Y., Notsu, K., Noguchi, M., Asada, T., (1980). Radon anomaly: A possible precursor of the 1978 Izu-Oshima-kinkai earthquake. *Science* 207, 882-883.

Yen, T. P., Chang, L. S., (1949). The Kwantzuling limestone. *Bulletin of the Geological Survey of Taiwan* 2, 51-52.

INDEX

D

G

H

I

J

K

L

M

N

O

P

Q

R

S

T

U

V

W

Y

Z